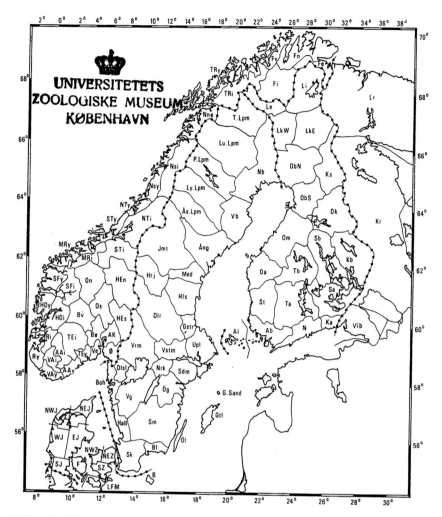

List of abbreviations for the provinces used throughout the text, on the map and in the following tables.

DENMARK

SJ	South Jutland	LFM	Lolland, Falster, Møn
EJ	East Jutland	SZ	South Zealand
WJ	West Jutland	NWZ	North West Zealand
NWJ	North West Jutland	NEZ	North East Zealand
NEJ	North East Jutland	B	Bornholm
F	Funen		

SWEDEN

Sk.	Skåne	Vrm.	Värmland
Bl.	Blekinge	Dlr.	Dalarna
Hall.	Halland	Gstr.	Gästrikland
Sm.	Småland	Hls.	Hälsingland
Öl.	Öland	Med.	Medelpad
Gtl.	Gotland	Hrj.	Härjedalen
G. Sand.	Gotska Sandön	Jmt.	Jämtland
Ög.	Östergötland	Ång.	Ångermanland
Vg.	Västergötland	Vb.	Västerbotten
Boh.	Bohuslän	Nb.	Norrbotten
Dlsl.	Dalsland	Äs. Lpm.	Åsele Lappmark
Nrk.	Närke	Ly. Lpm.	Lycksele Lappmark
Sdm.	Södermanland	P. Lpm.	Pite Lappmark
Upl.	Uppland	Lu. Lpm.	Lule Lappmark
Vstm.	Västmanland	T. Lpm.	Torne Lappmark

NORWAY

Ø	Østfold	HO	Hordaland
AK	Akershus	SF	Sogn og Fjordane
HE	Hedmark	MR	Møre og Romsdal
O	Opland	ST	Sør-Trøndelag
B	Buskerud	NT	Nord-Trøndelag
VE	Vestfold	Ns	southern Nordland
TE	Telemark	Nn	northern Nordland
AA	Aust-Agder	TR	Troms
VA	Vest-Agder	F	Finnmark
R	Rogaland		

n northern   s southern   ø eastern   v western   y outer   i inner

FINLAND

Al	Alandia	Kb	Karelia borealis
Ab	Regio aboensis	Om	Ostrobottnia media
N	Nylandia	Ok	Ostrobottnia kajanensis
Ka	Karelia australis	ObS	Ostrobottnia borealis, S part
St	Satakunta	ObN	Ostrobottnia borealis, N part
Ta	Tavastia australis	Ks	Kuusamo
Sa	Savonia australis	LkW	Lapponia kemensis, W part
Oa	Ostrobottnia australis	LkE	Lapponia kemensis, E part
Tb	Tavastia borealis	Li	Lapponia inarensis
Sb	Savonia borealis	Le	Lapponia enontekiensis

USSR

Vib	Regio Viburgensis	Kr	Karelia rossica	Lr	Lapponia rossica

FAUNA ENTOMOLOGICA SCANDINAVICA

Volume 15, part 1   1985

The Carabidae (Coleoptera) of Fennoscandia and Denmark

by

Carl H. Lindroth (†)

with the assistance of

F. Bangsholt, R. Baranowski, Terry L. Erwin, P. Jørum,
B.-O. Landin, D. Refseth and H. Silfverberg

E. J. Brill / Scandinavian Science Press Ltd.

Leiden · Copenhagen

Contents

8 colour plates are arranged after page 32.
List of references and index will be included with the second part of the volume.

Editors preface

Volume 15 of "Fauna ent. scand." will treat the Carabidae. The volume will appear in two parts. Each part will include a section with distribution tables, whereas the entire list of references and the index to the whole volume will be published with the second part.

The manuscript to the present volume was only partly ready when the author passed away in early 1979, and only with the assistance of many specialists it was possible to have it finished for publication. Dr. Terry L. Erwin (Washington, D. C., USA) revised the introductory chapters and the key to genera, and he also contributed with a new chapter introducing a modern system for the Nordic Caraboidea. Prof. B.-O. Landin (Lund, Sweden) revised and completed the section on the Bembidini.

The faunistic information for all four included countries have been greatly extended and revised compared to the original version, which lacked such information for many species. The new information is contributed by Mr. F. Bangsholt (Copenhagen, Denmark), Mr. R. Baranowski (Lund, Sweden), Mr. D. Refseth (Trondheim, Norway) and Dr. H. Silfverberg (Helsinki, Finland).

The sections on "Biology" under each species have been revised and greatly extended by Mr. P. Jørum (Viborg, Denmark). These sections were missing in the larger part of the manuscript, or they were ultra-short, only indicating the habitat preferences. Much new information is included, and many references to original modern papers have been included.

In the initial stage Mr. K. Arevad, Mr. M. Hansen and Mr. M. Holmen (all of Copenhagen, Denmark) gave valuable assistance with typing parts of Prof. Lindroth's handwritten manuscript.

Mr. O. Martin (Copenhagen, Denmark) arranged the specimens for the colour plates, and Mr. G. Brovad (Copenhagen, Denmark) prepared the photographic work with the plates.

All mentioned gentlemen are gratefully acknowledged for the careful and time-consuming work they have done.

Leif Lyneborg
Chief-editor

Introduction

This book may be regarded as an extension of my previously published revision of the ground-beetles of Sweden in the series "Svensk Insektfauna" (Lindroth 1942; 2nd ed. 1961). In this book all species of the three adjacent Scandinavian countries are included, Finland's frontiers being those recognized before 1940. The scope of the present work is geographically enlarged over that of previous editions only in so far as the entire concept "Fennoscandia orientalis" has been covered (as in *Catalogus Coleopterorum*, 1960), which has involved an addition of 9 species, known only from the Soviet Union. Since the ground-beetle faunas of adjacent parts of NW Europe have recently been treated by Lindroth (1974, Britain) and Freude (1976, Germany and surrounding countries), there is no need to include here descriptions of species restricted to these areas.

Descriptions of each species are kept rather short and I have concentrated on characters which provide ready separation from related species. The male genitalia are often described and illustrated, whereas the female genitalia, which often provide good characters, have been under-utilized.

Distributional data are given quite cursorily in the text. A summary of the occurrence of each species within different provinces throughout the area is presented as a catalogue at the end of each part of the volume. For detailed mapping, see Lindroth (1945b).

Information on life history and habits is given profusely; it may help the collector in finding specimens of a desired species, and it may also provide the incitement to further exploration of species in the field.

General Characteristics of Ground-beetle adults

The family Carabidae is here treated in its widest sense, that is including also the Tiger-beetles, which have often been regarded as a separate family, the Cicindelidae.

In almost all handbooks and catalogues of beetles, the Carabidae are at the beginning of the system. This does not necessarily mean that they constitute the most primitive forms among living beetles, i.e. that they represent more or less unchanged ancestors from which other existing families have evolved. In fact, the family Cupedidae, now absent from Europe, but occurring here in the early Oligocene when Baltic Amber was formed, is much closer to this position (see Crowson, 1955).

However, the Carabidae do represent a generalized type of beetle. Structurally, as running and predatory beetles, they could be regarded as non-specialized in comparison with the more advanced Polyphagan beetles. Consequently, carabid adults are distinguished in literature from other beetle families mainly by the "absence" of characters.

Together with the aquatic families Haliplidae, Dytiscidae, and Gyrinidae (in our fauna), as well as the terrestrial family Rhysodidae (by certain authors referred to the Carabidae), the Carabidae belong to the suborder Adephaga, characterized among other things by filiform antennae, five-segmented tarsi, coalescent basal segments of the abdomen, and the backwards produced meta-coxae (Fig. 2). The legs are slender, used for running or, in a few genera, the front pair for digging.

Other Coleoptera liable to be mistaken for Carabids are: (1) *Pteroloma*, a Silphid-beetle, originally described as a carabid; (2) certain members of the subfamily Omaliinae (Fam. Staphylinidae) with only slightly abbreviated elytra, but with a pair of ocelli on frons; (3) the genera *Crypticus* (Tenebrionidae) and *Anthicus* (Anthicidae), both with only 4-segmented meta-tarsi; (4) certain Cerambycids and Chrysomelids, with all tarsi (seemingly) 4-segmented.

External structure (Figs 1-6)

The carabid head capsule consists of several fused sclerites of which only the foremost, the clypeus (cly), is usually well-separated by a suture from the frons (fro); the frons, though, is not clearly delimited from the vertex. Behind the compound eyes (eye), the head is sometimes constricted, forming a neck. The underside of the head consists of the labium, divided into mentum (mnt) and gula (gul).

The movable appendages of the head are the antennae (ant), possessing 11 segments, and the mouth-parts: on the upper side, partly concealing the mandibles (mnd), is the labrum (lbr); below the mandibles are the complicated maxillae (max), each carrying a maxillary palp (mxp) and a segmented galea (gal) or "outer lobe." A pair of smaller labial palps (lbp) is fixed to the labium, which lies in front of the mentum. The anterior part of the labium carries an unpaired ligula (lig), or glossa, surrounded by a pair of paraglossae (par) ("ligula" is sometimes used for both organs together).

The upper side of the prothorax (prt) should rightly be termed the pronotum, as opposed to its lower surface, prosternum (prs), with its two lateral pro-episterna (pre). The wing-bearing meso- and meta-thorax are concealed under the elytra (ely), with the exception of the scutellum (scu), belonging to the mesothorax. On the underside (Figs 2, 6), the two segments are seen to consist of a central meso- and meta-sternum, respectively (mss, mts), each side bordering upon the corresponding episterna (mse, mte) to each of which usually one pair of small epimera (epm[1], epm[2]) are joined or fused.

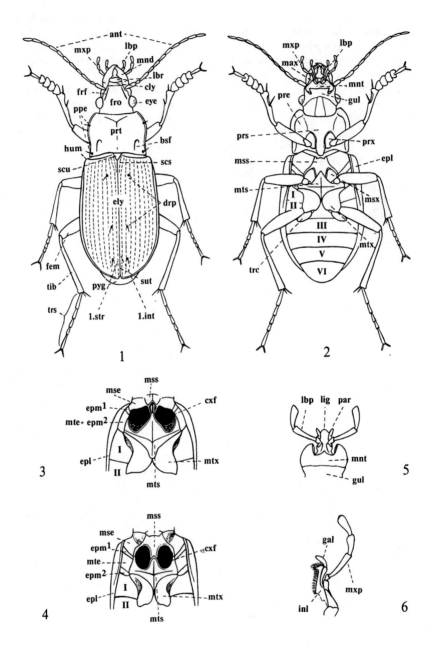

The elytra (ely), the forewings when in repose, meet along the suture (sut). Their lateral, reflexed part, not visible from above, are the epipleura (epl). The elytral striae and intervals, if present, are numbered from the centre to the lateral margin; the usually present abbreviated scutellar stria (scs), lies inside the first stria or between the first and second striae, and is not counted. Because these striae are often only rows of punctures, Erwin (1974) introduced the term "interneur" for the structure, which then may take the form of a stria, a sulcus, or a row of punctures. Additional dorsal punctures (drp), with attached seta, are often present, usually on the third interval or touching adjoining striae. The hind-wings, if fully developed, have a reflexed apical part. Their venation undoubtedly possesses taxonomically useful characters, but it has not been used in this book.

The abdomen is composed of segments, terga on the upper, sterna on the lower side. Only six sterna (I-VI) are visible (except in *Brachinus*), the foremost sternum is only visible laterally. The last tergum, if visible, is called the pygidium (pyg).

The innermost part of each leg is the coxa (prx, msx, mtx), to which the femur (fem) and the trochanter (trc) are attached. Then follow the tibia (tib) and the five-segmented tarsus (trs) with a pair of claws in terminal position on last tarsomere.

Figs 1-6. General structure of a ground-beetle (Carabidae). 1: upper side; 2: lower side; 3: meso- and metathorax in *Carabus;* 4: same in *Pterostichus;* 5: labium; 6: maxilla.

ant	– antenna	mse	– mes-episternum
bsf	– basal fovea of prothorax	mss	– mesosternum
cly	– clypeus	msx	– meso-coxa
cxf	– meso-coxal cavity	mte	– met-episternum
drp	– dorsal punctures	mts	– metasternum
ely	– elytra	mtx	– meta-coxa
epl	– epipleura of elytra	mxp	– maxillary palp
epm[1]–	epimeron of mesosternum	par	– paraglossa
epm[2]–	epimeron of metasternum	ppe	– setae of prothorax
eye	– compound eye	pre	– pro-episterna
fem	– femur	prs	– prosternum
frf	– frontal furrow	prt	– prothorax
fro	– frons	prx	– pro-coxa
gal	– galea	pyg	– pygidium
gul	– gula	scs	– scutellar stria
hum –	.humerus	scu	– scutellum
inl	– inner lobe of maxilla	sut	– suture of elytra
lbp	– labial palp	tib	– tibia
lbr	– labrum	trc	– trochanter
lig	– ligula	trs	– tarsus
max –	maxilla	1.int	– first elytral interval
mnd –	mandible	1.str	– first elytral stria
mnt –	mentum	I-VI	– abdominal sternites

13

Larval Characteristics

Carabid larvae are of the "campodeid" type (except in *Cicindela,* and later stages of the ectoparasitic genera, *Lebia* and *Brachinus*). The normal larvae are slender, long-legged and have well developed cerci on the ninth abdominal segment. From staphylinid larvae, they are distinguished by the presence of six segments on each leg, the claw(s) not being fused to the tarsus.

In general, the larvae are pronouncedly predatory, more so than the adults, but their concealed mode of life and predominantly nocturnal habits have hampered a thorough study of their taxonomy and natural history. Much remains to be done in this field, notably by rearing from gravid females.

The larvae are not mentioned here, since the late Dr. S. G. Larsson and Dr. Martin Luff will produce a special volume on the subject.

The System of the Carabidae

The family Carabidae is generally divided into several subfamilies, though their arrangement and content have been much varying between specialists. In Europe, the classification introduced by Ganglbauer (1892) and largely based on that of Leconte & Horn (1883) in North America, was long prevailing. Major changes were proposed by Jeannel (1941-42), who split the Carabidae *s. lat.* into many new families; but his system, though containing many excellent ideas, has not been generally accepted.

A slightly modified Leconte-Horn-Ganglbauer system was introduced for the North American fauna (Ball, 1960; Lindroth, 1969). Applied to Scandinavian groups, it implies the recognition of only five subfamilies: Cicindelinae, Trachypachinae, Omophroninae, Carabinae, and Brachininae. All other suprageneric groups were regarded as tribes (with subtribes) and the traditional limit between the Carabinae (in its restricted sense) and the Harpalinae, as well as the creation of an intervening subfamily, the Scaritinae (Crowson, 1955), were regarded as artificial. More recently (Erwin, 1979, 1983; Kryshanovskij, 1976) refined this classification, giving family status to the Trachypachidae, and only recognizing three carabid subfamilies, namely the Omophroninae, Paussinae, and Carabinae. They also use numerous supertribes, striking a balance between the Jeannelian concept and the "North American" one of Ball and Lindroth. This last system is reviewed in a following chapter because it reflects extensive research done over the last twenty years on Carabidae.

Nomenclature

In Lindroth (1963-1969), I prefered to use taxonomic concepts on a higher level (subfamilies, tribes, etc.) and in the widest possible sense — I am a "lumper". This applies also to generic names. Their purpose, in a faunal treatment where all species of a genus are not revised, is not only to indicate some sort of a relationship between species, but also, from a purely practical point of view, to give even the non-specialist an idea of where the species belongs. Due consideration to proposed relationships may be paid by arranging the species into "natural groups", at least within large genera which then can be named informally, thus according to general European practice, subgeneric names have been used for this purpose.

It is very important, in the interest of stability and continuity, that changes of Latin names, notably at the generic and specific levels, are restricted to a necessary minimum. The *International Code of Zoological Nomenclature* (1961), fortunately, made provision for suspension of the priority principle, which had been followed too rigidly by many authors.

Quite recently, Silfverberg (1976) found several homonyms among Fennoscandian species of Carabidae originally described under the collective generic name "Carabus" (e.g. *Carabus bipustulatus* Fabricius, 1775 = *Panagaeus; Carabus bipustulatus* Fabricius, 1792 = *Badister*) and consequently changed the species name of the younger homonym.

Infra-specific nomenclature is a far more intricate and not purely formal problem. On no level of taxonomy has the creation of new names been performed with less skill than by introduction of subspecific names. They are justified only if based on good quantities of statistically treated material. Many subspecies are artificial, in that the variation within the species could better be described as one or more "clines," that is, geographical zones of gradation with respect to a single character. In this book, subspecific names have been used with care. Variations of lower value (variatio, aberratio, etc.) are named only in the interest of synonymy, thus ridding nomenclature of unnecessary names.

Brief Review of the Literature on Scandinavian Carabidae

As everywhere, no lasting work on the carabid fauna of Scandinavia could start until the binomial nomenclature was introduced by Linnaeus (1758). He himself described 61 (or 62) species, most of them in the genus *Carabus* and these are still valid, though generally referred to other genera (Lindroth, 1957).

The foremost successor of Linnaeus, as an entomologist, was the Dane J. C. Fabricius, who, in his works between 1775 and 1801, described a vast number of new species, of which almost 300 belonged to the Carabidae (including many exotics). They were placed under eleven different generic names, but an overwhelming majority (more than 200 species) were still referred to *Carabus*.

The first local faunas to appear were G. Paykull's "Monographia Caraborum Sueciae" (1790) and "Fauna Suecica" (1798), which did not show much progress in description and classification beyond Fabricius; the Carabidae were divided into only 6 genera. Far more important was L. Gyllenhal's "Insecta Suecica" (4 parts, 1808-27), with all species described in detail, many of them new, and the Carabidae (1810, 1813) arranged in 19 genera, many adopted from Latreille in France. The first coleopterous fauna of Finland, the "Insecta Fennica" (1817-34) by C. R. Sahlberg, was mainly based on Gyllenhal.

The leading Danish entomologist of the century, J. C. Schiødte, published the important "Genera og Species af Danmarks Eleutherata" in 1841, where only the Adephaga were treated. Schiødte most lasting achievement, however, was the long series (1861-83) of still unsurpassed larval descriptions, including many Carabidae, appearing under the title "De Metamorphosi Eleutheratorum".

A contemporary of Schiødte was C. G. Thomson in Sweden, who devoted the first half of his life to Coleoptera and in his grand work "Skandinaviens Coleoptera" (10 parts, 1859-63) introduced a partly new classification also for the Carabidae (1859).

The exploration of Norway, by J. Sparre Schneider and T. Munster, was not seriously begun until the last decades before 1900. During the 20th century, most Scandinavian carabidologists have been interested in revising small taxonomic units, as well as in natural history, experimental ecology, and zoogeography.

It must be realized, however, that the composition of a fauna is never stable. Several carabid species have been threatened or may even become extinct, due to human intrusion upon their habitats. Other species, in a state of expansion, have invaded our area during the present century.

Distribution and State of Knowledge

Since carabid beetles have always been among the favourite objects of collectors — and also because they constitute the beginning family in almost all books on Coleoptera — they are well represented in public and private collections and our knowledge of even the detailed distribution of the species within our area must be regarded as satisfactory. These were mapped by Lindroth (1945b), and with regard to Denmark in detail by Bangsholt (1983).

The approximate number of known species (Fig. 7), as well as an estimate of the relative exploration of Carabidae, in different parts of Fennoscandia were published in Lindroth (1949). The values may not have changed much since, but it is important to keep in mind, that true faunal changes have taken place — and still take place — mainly through immigration from outside and in part favoured by human transformation of the landscape into open country (Lindroth, 1972). A unique case of rapid immigration among the Carabidae is *Amara majuscula* Chaudoir.

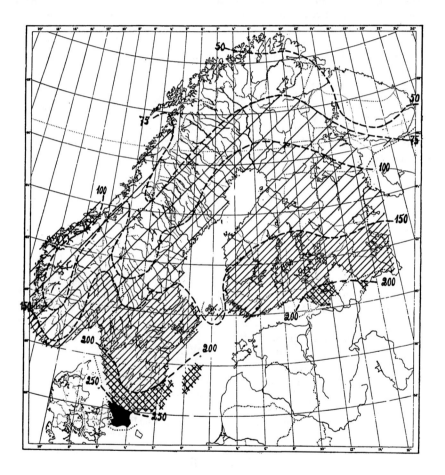

Fig. 7. Approximate number of Carabid species in different parts of Fennoscandia. (After Lindroth 1949). For details on Danish figures, see Bangsholt (1983: Fig. 3).

17

Killing and Mounting

The best method of killing beetles is by ethyl acetate (acetic ether). This substance keeps the specimens soft and relaxed for months, ready for immediate mounting and dissection. Since beetles collected in different localities and habitats should always be kept apart during collecting, it is advisable to bring a good supply of glass tubes (or plastic vials), each containing strips of filter paper or a portion of coarse hardwood sawdust (poplar is preferred) moistened, but not dripping wet, with ethyl acetate.

Cyanide and/or strong alcohol, generally used in the past, especially among North American collectors, make the insects stiff, difficult to mount and dissect afterwards, and in some species cause distortion of the internal sac of the penis, with dislocation of its taxonomically important structures as a consequence.

Large specimens may be pinned directly through the basal part of the right elytron, but the majority should be glued to a piece of cardboard; either to the tip of a small triangular point or on a rectangular mount of a larger size than the insect. The first method has the advantage of easier examination of the underside, but the insect is better protected on a rectangular mount and, in genera where the characters of the ventral side are important, one or two specimens of a series may be mounted upside down. Many different kinds of glue are in use (water-soluble fish glue is excellent); it is important, if antennae and legs are spread and fixed in a meticulous way, that the upper surface of body and appendages is not brushed with glue so as to conceal pubescence and other subtle structures.

Rearing

This is a fascinating, but little practiced task. The most reliable method is to obtain *ex ovo* larvae from pregnant females kept in captivity with soil from the habitat. The first instar larva possesses important taxonomic characters and some specimens of this stage should be killed and preserved (in 80% alcohol). During subsequent development, it is wise to keep each larva in a separate tube, because they are very often cannibals. The substrate should be kept too dry rather than too wet. Small cut pieces of live earthworms are usually sufficient food; even dog biscuits can be used.

Unknown larvae found in the field are of course best identified be rearing them to adult beetles. It is important to save the last cast larval skin lying at the apex of the pupa; it retains all essential morphological features characterizing the larva.

Natural History

In older works, the Adephaga (Carabidae and related aquatic families) were referred to as "Carnivora", thus implying a predatory mode of life. This opinion is still generally stressed in current entomological, particularly the applied, literature. The entire habitus of most ground-beetles is undeniably that of a predator, but careful studies of their feeding habits in recent years have revealed that many species, perhaps the majority, have a mixed diet and should rather be termed "omnivorous". Members of such genera as *Amara* and *Harpalus* no doubt feed more on seeds and other vegetable matter than on animals. Carabid larvae, however, often seem to be more pronounced carnivores.

Specialized predators are few. They include species of *Calosoma*, the "Caterpillar Hunters"; *Cychrus* and *Licinus* on shell-bearing snails; most *Dyschirius* on staphylinids of the genus *Bledius*. Larvae of *Brachinus* and *Lebia* are ectoparasites on certain beetle pupae.

In northern countries, carabids are usually pronounced ground-dwellers — from which they got their English name. Towards the Equator, planticolous (including arboricolous) forms increase in number, but in our area only two species of *Calosoma*, two of *Agonum*, several species of *Dromius*, and *Tachyta nana*, are regular tree-climbers. Other species, for instance of the genera *Harpalus* and *Amara*, climb herbaceous plants at night in search of vegetable food.

Most Carabidae are long-lived in the adult stage (*Carabus* and other large species normally live more than two years). This is fortunate for the collector, in that it allows a fairly exhaustive exploration of a resticted area within a short period of time, anytime of the year. It is, however, necessary in this context to distinguish between larval hibernators and those, the majority, which over-winter in the adult stage. The former (e.g. many *Amara* species) have their peak of abundance in mid-summer and are often not found before June. Adult hibernators, on the other hand, are usually scarce in the middle of the summer, the time of larval development.

The condition of the hind-wings is subject to wide variation in the Carabidae. In most species, the wings are fully developed, but only a few *(Cicindela, Bembidion* subg. *Chrysobracteon)* use them regularly for predation and escape. The main purpose of flight is to support migration into new habitats, notably between winter and summer quarters. In constantly flightless species such as most *Carabus,* the wings are usually reduced to a tiny rudiment and the elytra may be fused together along the suture (e.g. *Cychrus, Pterostichus* subg. *Cryobius*). Wing dimorphism is frequent, that is, long-winged (macropterous) and short-winged (brachypterous) individuals are present in the same species, regardless of sex, and populations are usually mixed in this respect. It is, however, important to emphasize that all long-winged individuals and species are not necessarily able to fly, as the flight muscles may be reduced and non-functional.

The sections on "Biology" under the various species contain numerous references to papers dealing with biological matters. An important reference to further reading on Carabid ecology and biology is Thiele (1977).

Collecting

The easiest way to collect Carabidae is by turning over stones. But even in a stony field many species prefer other micro-habitats. It is also always rewarding to look for them under depressed mats of vegetation, such as *Calluna* and *Empetrum,* under the leaf rosettes of *Artemisia, Rumex,* and thistles, etc. Some species bury themselves rather deeply in the soil and may be discovered by pulling up clumps of tall plants and shaking the roots over a piece of cloth or paper. For extracting small species living in the leaf-litter under trees and bushes, in heaps of grass and other vegetables, or in not too wet moss, the ordinary insect sieve is indispensable; it is the most efficient tool for collecting insects hibernating in the soil. Leaf-litter and flood refuse on the sea shore and the banks of lakes and rivers may also be thrown into the water, so that the inhabitants are forced to surface and thus are easily caught.

Special methods are required for collecting in moist localities. Soft mats of vegetation at the margin of lakes and pools may often be submerged entirely by treading them down into the water and the floating insects are rapidly apparent. "Treading" is also commendable in *Carex* and moss vegetation on somewhat firmer soil, for instance in *Sphagnum* bogs. On banks and shores with sparse or no vegetation most beetles are concealed under the surface (e.g. *Dyschirius*). Most are immediately exposed if the habitat is profusely splashed with water; some species require more patience of the collector and will only surface after a few minutes.

A convenient method of collecting all kinds of beetles running on the surface of the ground, especially at night, and indispensable to students interested in quantitative population investigations, is the pit fall trap which works automatically 24 hours a day. In firm soil, it is sufficient to dig holes with perpendicular walls; in other places flower-pots or glass jars with the upper margin buried to the level of the ground surface may be used. Left alive in the trap, even for a few hours only, carabids will mutilate each other; they may also be picked up by birds. It is therefore better to let them fall into some killing and preserving fluid, such as formalin (ca. 4%) or ethylene glycol. A few drops of detergent added to these liquids will lower the surface tension and the insects will immediately sink to the bottom, unable to escape.

Labelling

It is of utmost importance to develop a strict method of labelling field collections and the individual insects in the course of their handling towards final deposition in private or museum collections. Any risk of mis-labelling must be avoided and it is quite detestable to rely upon ones memory alone.

In the first place, field collections from different geographical localities and different habitats must be kept apart from the very beginning in separate vials. Each vial must immediately be provided with a provisional hand-written label with notes describing the geographical locality, condensed habitat description, and date. A number alone is never sufficient, but is conveniently added as a reference to a simultaneously kept journal, where the different field collections with their environment are more carefully described, each under its number in chronological order.

It is only exceptionally feasable to make direct measurings of abiotic factors, such as temperature, humidity, acidity, on the spot. But sun-exposure, soil conditions, etc., may be noted; and, above all, the best and most easily conceived influence of abiotic factors upon the habitat is manifested in the vegetation. Therefore, under the relevant number in the journal, information on at least the most predominant plant species of the habitat is taken. This means, that a naturalist without sufficient botanical knowledge will never become a "first class" collector of carabid beetles.

After mounting all specimens from one catch, they may provisionally be kept together in a box, with the original label on the first specimen awaiting final, preferably printed, labels to be produced. The following content of these labels is: (a) geographical definition (country or province, locality, if possible pin-pointed according to generally accepted grid-net); (b) habitat description (e.g. by reference to a journal number); (c) date; (d) collector.

Notes on identification

Many large species of Carabidae or even members of some small genera are easy to identify in the field, either with the naked eye or with the aid of a hand-lens (10-20X). Quite the contrary is the case in large genera such as *Dyschirius, Bembidion, Agonum, Amara, Harpalus,* etc., in which certain species can be reliably named only after investigation of the male genitalia, as described below.

A study of the microsculpture of the upper surface often is very helpful and may be unavoidable in some cases, particularly in the females (which as a rule are more coarsely sculptured). If present, the microsculpture usually consists of coherent lines which either join into meshes, from isodiametric to very transverse, or run very close together in a parallel arrangement, producing a more or less pronounced iridescent lustre, notably on the elytra (Lindroth, 1974). An investigation of the microsculpture requires a magnification of about 100X and strong light. Best for the purpose is the so-called "ultropaque" with a lamp built into the tube, or of course an electronic "scanning" microscope. But these are expensive and as a substitute the use of an ordinary compound microscope, with sideways florescent light, is recommended.

Preparations of male genitalia (Fig. 8). These, more or less markedly sclerotized structures, consist of penis (median lobe) and two parameres (lateral lobes), which together form the aedeagus. In fresh or softened specimens, these organs are easily extracted without any visible damage ensuing. Dissection is best made under a microscope or a strong fixed lens. The beetle is placed in water on a glass and the male organ extruded, using ordinary sharp insect pins (Fig. 8). The aedeagus, still in water, is cleaned from ligaments and muscles under high magnification. If the species investigated is sufficiently characterized by the outer structures of penis and parameres, the organ, after study, is simply glued to a piece of cardboard on the same pin as the beetle. Frequently, however, the internal structures of the penis are decisive and the organ has to be made transparent. After about 12 hours in cold caustic potash (15% KOH), it is carefully washed in water and then in absolute alcohol and finally transferred to at drop of clove-oil. After varying time (one day or more), the clearing process is finished, a coverglass is added, and every detail of the armature and foldings of the internal sac can be investigated through a compound microscope. The aedeagus may be stored in clove-oil for months, but eventually becomes very brittle. After study, the organ may either be preserved, directly transferred from clove-oil, as a permanent canada-balsam slide between glasses; or, after passing through absolute alcohol and water, it may be dried and glued on card and attached to the same pin as the specimen. The latter procedure is perhaps preferable, because it eliminates any risk of the genital slide being permanently separated from the beetle.

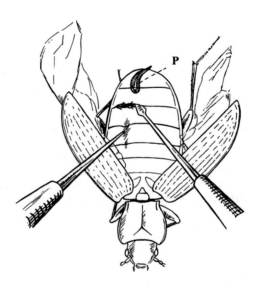

Fig. 8. Dissection of male genitalia. P – penis (with parameres).

A New Classification
of Caraboidea of Fennoscandia

(from Erwin & Sims 1984)

SUPERFAMILY CARABOIDEA

I. FAMILY TRACHYPACHIDAE
    1. TRIBE Trachypachini

II. FAMILY CARABIDAE
  A. SUBFAMILY CARABINAE
    a. SUPERTRIBE Nebriitae
      2. TRIBE Nebriini
      3. TRIBE Notiophilini
    b. SUPERTRIBE Loriceritae
      4. TRIBE Loricerini
    c. SUPERTRIBE Carabitae
      5. TRIBE Carabini
      6. TRIBE Cychrini
    d. SUPERTRIBE Cicindelitae
      7. TRIBE Cicindelini
    e. SUPERTRIBE Omophronitae
      8. TRIBE Omophronini

  B. SUBFAMILY SCARITINAE
    f. SUPERTRIBE Elaphritae
      9. TRIBE Elaphrini
    g. SUPERTRIBE Scarititae
      10. TRIBE Clivinini

  C. SUBFAMILY PAUSSINAE
    h. SUPERTRIBE Brachinitae
      11. TRIBE Brachinini

  D. SUBFAMILY BROSCINAE
    i. SUPERTRIBE Broscitae
      12. TRIBE Broscini

  E. SUBFAMILY PSYDRINAE
    j. SUPERTRIBE Psydritae
      13. TRIBE Patrobini
    k. SUPERTRIBE Rhysoditae
      14. TRIBE Rhysodini
    l. SUPERTRIBE Trechitae
      15. TRIBE Trechini
      16. TRIBE Pogonini
      17. TRIBE Bembidiini

  F. SUBFAMILY HARPALINAE
    m. SUPERTRIBE Pterostichitae
      18. TRIBE Pterostichini
      19. TRIBE Zabrini
    n. SUPERTRIBE Panagaeitae
      20. TRIBE Panagaeini
    o. SUPERTRIBE Callistitae
      21. TRIBE Callistini
      22. TRIBE Oodini
      23. TRIBE Licinini
    p. SUPERTRIBE Harpalitae
      24. TRIBE Harpalini
    q. SUPERTRIBE Lebiitae
      25. TRIBE Perigonini
      26. TRIBE Masoreini
      27. TRIBE Odacanthini
      28. TRIBE Lebiini

Key to Genera of Fennoscandian Carabidae

1   Entire elytron uniformly pubescent, or at least with one row of setae or bristles along entire length of each (or every second) interval* 2

–   Elytron glabrous, except for marginal setae and often setiferous "dorsal" punctures on intervals 2-3, or with only outer intervals pubescent 18

2(1)   Elytron with well developed interneurs, either striate or punctulate rows (at least on the median parts) 3

–   Elytron without regular interneurs, (though sometimes with a few shallow, impunctate furrows) 16

3(2)   Elytron with apex transversely truncate, leaving at least last abdominal tergum visible (Fig. 10) 4

–   Elytron with apex rounded, last tergum quite or almost concealed (except in gravid females) 6

4(3)   Elytra bright metallic, blue or green (Pl. 8: 6,7)..........................
.... *Lebia* Latreille *(cyanocephala)*

–   Entire body unmetallic 5

5(4)   Pronotum glabrous; 4th tarsal segment dilated (Fig. 42) ....
.... *Demetrias* Bonelli *(atricapillus)*

–   Pronotum pubescent; 4th tarsal segment normal (Pl. 8: 18) ....
.... *Cymindis* Latreille

6(3)   Frontal furrows sharp, semicircularly diverging behind eyes. Small species (not over 5.5 mm.)... 7

–   Frontal furrows not prolonged, often obsolete. Usually larger 8

7(6)   Less than 3 mm. Terminal segment of maxillary palp narrow (Fig. 13) (Pl. 4: 17) .... *Perileptus* Schaum

–   More than 4 mm. Terminal segment of maxillary palp normal (Fig. 12) (Pl. 5: 2,3) .... *Trechus* Clairville *(discus* or *micros)*

8(6)   Head with very narrow neck (Fig. 9). Elytra rufous with black cross. (Pl. 8: 3) .... *Panagaeus* Latreille

–   Head without pronounced neck. Elytra otherwise marked.............. 9

9(8)   Elytron without abbreviated scutellar stria. Not above 7.5 mm 10

–   Elytron with abbreviated scutellar stria. Usually larger 12

10(9)   Sutural stria of elytron recurrent at apex along outer margin (as in Fig. 24). Intervals with very short, scattered setae (Pl. 5: 22) .... *Tachyta* Kirby

–   Sutural stria not recurrent. Setae on intervals evident 11

---

* Seta-bearing punctures (mainly on head and pronotum) are often referred to; if the setae is broken, its fix-base can be seen as a pupillate puncture.

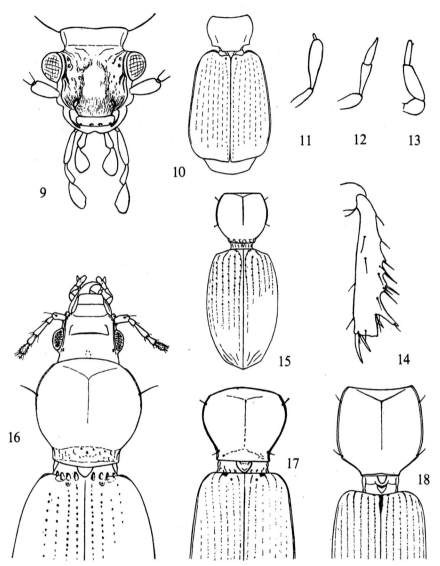

Figs 9-18. 9: head of *Panagaeus cruxmajor* (L.); 10: pronotum and hindbody of *Lebia cyanocephala* (L.); 11: maxillary palp of *Bembidion;* 12: same of *Trechus;* 13: same of *Perileptus;* 14: mesotibia of *Clivina collaris* (Hbst.); 15: pronotum and elytra of *Dyschirius luedersi* Wagn.; 16: *Miscodera arctica* (Payk.); 17: pronotum and part of hindbody of *Broscus cephalotes* (L.); 18: pronotum and elytra of *Clivina fossor* (L.).

11(10) Elytron with a single row of punctures and erect setae on
each interval. Legs with only tibiae in part pale . . . .
. . . . *Trichocellus* Ganglbauer *(mannerheimi)*
  –   Elytral intervals irregularly punctate. On the legs at least
tarsi pale. (Pl. 7: 5) . . . . *Dicheirotrichus* Duval
12(8) Upper surface of tarsi glabrous (except for terminal setae) 13
  –   Upper surface of tarsi pubescent 14
13(12) Antennae with 3 glabrous basal segments (except for the
usual terminal setae). At least 8.5 mm. (Pl. 7: 17-19) . . . . . *Chlaenius* Bonelli
  –   Antennae pubescent from 3rd segment, usually 6-7 mm . . . .
. . . . *Harpalus* Latreille *(signaticornis)*
14(12) Elytra with sharp bicoloured pattern. (Pl. 7: 3) . . . . . *Diachromus* Erichson
  –   Elytra unicolorous, dark (or slightly paler along the suture) 15
15(14) Frons and temples with decumbent, and dense, pube-
scence . . . .*Harpalus* Latreille (subg. *Ophonus* Dejean)
  –   Head (except for supra-orbital setae) glabrous. (Pl. 3: 12)
. . . . *Harpalus* Latreille (subg. *Pseudoophonus* Motschulsky)
16(2) Elytron with rounded apex (Fig. 145). Terminal segment of
maxillary palpi rudimentary (as in Fig. 11). (Pl. 5: 5) . *Asaphidion* des Gozis
  –   Elytron truncate at apex. Terminal segment of maxillary palpi normal . . 17
17(16) Not over 2.4 mm. Body testaceous. Eyes very small (Fig.
131). (Pl. 5: 1) . . . . *Aepus* Samouelle
  –   More than 6 mm. Body bicoloured (elytron dark). Eyes normal, protruding
(Pl. 8: 19) . . . . *Brachinus* Weber
18(1) Mesonotum (with extreme base of elytron) strongly con-
stricted as a "peduncle", upon which the scutellum is situ-
ated (Figs 15-18) 19
  –   Body not pedunculate 22
19(18) 6.5 mm or more; 3rd antennal segment twice a long as 2nd 20
  –   6.5 mm or less; 3rd antennal segment shorter than 2nd 21
20(19) 16 mm or more; pronotum with 2 lateral setae (Fig. 17).
(Pl. 3: 3). . . . . *Broscus* Panzer
  –   8 mm or less; pronotum only with anterior lateral seta
(Fig. 16). (Pl. 4: 15) . . . . *Miscodera* Eschscholtz
21(19) Lateral bead of pronotum prolonged behind posterior seta
(Fig. 14). (Pl. 4: 13) . . . . *Clivina* Latreille
  –   Lateral bead of pronotum not prolonged (Fig. 15). Meso-
tibia without spine. (Pl. 4: 14) . . . . *Dyschirius* Panzer
22(18) Elytron with 11, or more, well impressed, at least basally
regular striae, but without ridges or tubercles 23
  –   Elytron with less than 11 striae (abbreviated scutellar stria
not counted) or without regular striae 25
23(22) Scutellum concealed under median lobe of pronotum. Pro-
sternum covering mesosternum (Fig. 55). (Pl. 4: 2) . . . . *Omophron* Latreille

26

–	Scutellum visible. Mesosternum not concealed 24
24(23)	16 mm or more. Neck not constricted. Antennal setae normal (Fig. 58). (Pl. 1: 4-8) .... *Calosoma* Weber
–	Under 9 mm. Neck strongly constricted. Antennal segments 2-4 with long setae (Fig. 95). (Pl. 4: 12) .... *Loricera* Latreille
25(22)	Head with clypeus broader than distance between antennae (Fig. 44). Elytra without striae, dark with pale spots (Figs 47-50). (Pl. 1: 1-3) .... *Cicindela* Linnaeus
–	Clypeus narrower than distance between antennae 26
26(25)	Maxillary palpi with last segment rudimentary (Fig. 11) (Small species, not more than 7.5 mm) 27
–	Maxillary palpi with well developed terminal segment (at least as in Fig. 12) 29
27(26)	Elytron with abbreviated scutellar stria. Pro-tibia truncate at apex. (Pl. 5: 6-20) .... *Bembidion* Latreille
–	Elytron without scutellar stria. Pro-tibia with oblique apex (Fig. 245) ... 28
28(27)	Sutural stria connected at apex in a wide bow with one of the outer striae (Fig. 24). (Pl. 5: 21).................... *Tachys* Dejean
–	The recurrent part of the sutural stria running parallel with side-margin (Pl. 5: 22) .... *Tachyta* Kirby
29(26)	All antennal segments, except for terminal setae, without pubescence (Fig. 54). Metacoxa extended laterally to meet the margin of elytron (Fig. 53). (Pl. 4: 1) .. *Trachypachus* Motschulsky
–	At least the last 7 segments of antennae with dense, depressed pubescence. Metacoxa not extended laterally 30
30(29)	Frons with 6 sharp longitudinal carinae (Fig. 27). 2nd elytral interval much broader than all following (Figs 85, 86). (Pl. 4: 7)............... : .. .. *Notiophilus* Duméril
–	Frons without carinae (or with 4, or less, furrows). 2nd elytral interval not outstanding in width 31
31(30)	Elytron each with 3 rows of ocellate depressions but without or with highly disturbed striae (Fig. 92). (Pl. 4: 10) . *Elaphrus* Fabricius
–	Elytral sculpture otherwise 32
32(31)	Sutural stria of elytron "recurrent" at apex (Fig. 24). Frontal furrows prolonged and semicircularly diverging behind eyes (Not over 6.5 mm) 33
–	Sutural stria not recurrent. Frontal furrows not or less prolonged 34
33(32)	Eyes rudimentary (Fig. 131). Elytron with sparse erect hairs. Under 2.5 mm. (Pl. 5: 1) .... *Aepus* Samouelle
–	Eyes normal (Fig. 132). Elytron except for dorsal punctures, glabrous. Not below 3.5 mm. (Pl. 5: 2-4) .... *Trechus* Clairville
34(32)	Frons with a single seta-bearing "supra-orbital" puncture inside eye 35
–	Frons with at least 2 supra-orbital punctures, the posterior

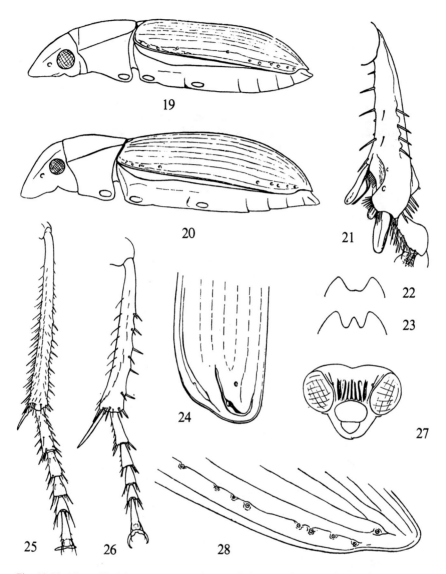

Figs 19-28. 19: profile of body of *Acupalpus;* 20: same of *Bradycellus;* 21: left front-tibia of *Zabrus tenebrioides* (Gz.); 22: mentum of *Olisthopus;* 23: same of *Agonum;* 24: elytron with recurrent 1st stria of *Trechus;* 25: hind leg of *Anisodactylus binotatus* (F.); 26: same of *Harpalus tardus;* 27: head of *Notiophilus palustris* (Dft.); 28: posterior group of marginal elytral punctures in *Stenolophus.*

only. (Species above 5.2 mm with strongly iridescent elytra) 48

47(46) First segment of metatarsus not longer than terminal spur
of tibia (Fig. 26). Elytron usually with humeral tooth. (Pl.
3: 12 & 7: 1,2). .... *Harpalus* Latreille

–   First segment of metatarsus longer than terminal spur (Fig.
25). Elytron without humeral tooth. (Pl. 7: 4) ...... *Anisodactylus* Dejean

48(46) Antennae entirely pale. Mentum with median tooth (as in
Fig. 23). Elytron without coherent microsculpture, not
iridescent. Body convex (Fig. 20). (Pl. 7: 7,8) .... *Bradycellus* Erichson

–   Antennae dark with pale base. Mentum without tooth
(Fig. 22). Elytra more or less iridescent from transverse microsculpture (except in *Acupalpus meridianus*). Body flatter (Fig. 19) 49

49(48) 5 mm or more. The row of marginal elytral punctures with
pronounced gap posteriorly (Fig. 28). (Pl. 7: 9, 10).... *Stenolophus* Dejean

–   4.5 mm or less. Marginal row of elytral punctures more
continuous (as in Fig. 19). (Pl. 7: 11, 12) .... *Acupalpus* Latreille

50(34) Elytral apex rounded or sinuate, in normal position covering entire abdomen or leaving only a lesser part of last
segment free (notably in gravid females) 51

–   Apex of elytron transversely or obliquely truncate, leaving
at least most of terminal segment uncovered 72

51(50) 1st antennal segment at least 3 times as long as 2nd 52

–   1st antennal segment not more than twice as long as 2nd 54

52(51) Elytron without abbreviated scutellar stria. Mandibles very
long (Fig. 29). (Pl. 6: 2).............................. *Stomis* Clairville

–   Elytron with scutellar stria. Mandible short and broad 53

53(52) Elytral intervals with coarse punctuation. (Pl. 8: 1) .... *Licinus* Latreille

–   Elytral intervals impunctate. (Pl. 7: 14-16) .... *Badister* Clairville

54(51) Base of pronotum with sharp lateral incision (Fig. 35).
3-4 mm. (Pl. 8: 16) .... *Lionychus* Wissmann

–   Pronotum without or with shallow incision laterally................. 55

55(54) Head and pronotum with strong, evenly distributed punctuation. (Figs 88, 89). (Pl. 4: 9) .... *Diacheila* Motschulsky

–   Head and pronotum at least medially impunctate 56

56(55) Frons inside each eye with double, parallel or slightly converging furrows (Fig. 32) 57

–   Frons without or with obscurely delimited, never doubled,
furrows or foveae 59

57(56) Frontal furrows each side joined by a transverse groove
(Fig. 87). Elytron with two rows of large foveae. (Pl. 4:8) .. *Blethisa* Bonelli

–   Head without transverse groove. Dorsal punctures of elytron not foveate 58

58(57) Elytral base margined almost to scutellum. Upper surface

30

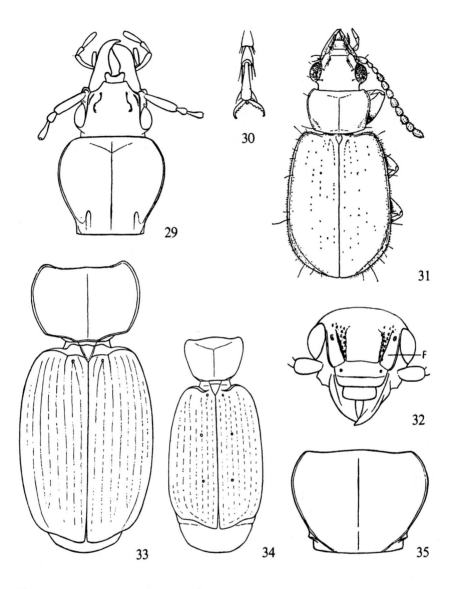

Figs 29-35. 29: head and pronotum of *Stomis pumicatus* (Pz.); 30: terminal tarsal segments of *Synuchus vivalis* (Il.); 31: *Perigona nigriceps* (Dej.); 32: head of *Patrobus atrorufus* (Strøm); 33: pronotum and hindbody of *Masoreus wetterhalli* (Gyll.); 34: pronotum and hindbody of *Syntomus;* 35: pronotum of *Lionychus quadrillum* (Dft.).

more or less metallic. (Pl. 6:1) .... *Pogonus* Dejean
— Elytron not margined inside shoulder. No metallic lustre
  (Pl. 4: 16) .... *Patrobus* Stephens
59(56) Elytron with obsolete striae, except stria 8, which is deepened apicad. (Fig. 31) 2.0-2.5 mm. (Pl. 7: 13) .... *Perigona* Castelnau
— Elytral striae normal, though sometimes faint (stria 9
  deepened apicad in *Abax*). 60
60(59) Pronotum with base sinuate laterally (Figs 33, 34). Elytra
  leaving at least apical half of last tergite free. (Under 6 mm) 61
— Base of pronotum not or very faintly sinuate inside hind
  angles. Last tergum covered by elytra (except in gravid females) 62
61(60) Elytron with pale base. Pronotum almost as broad as elytra (Fig. 33). (Pl. 8: 5) .... *Masoreus* Dejean
— Elytron uniformly dark. Pronotum much narrower (Fig.
  34). (Pl. 8: 15) .... *Metabletus* Schmidt-Göbel
62(60) Claws denticulate or pectinate internally, at least at base. (Fig. 30) 63
— Claws smooth 66
63(62) All tarsi pubescent above. Upper surface with metallic
  lustre. 13 mm og more .... *Laemostenus* Bonelli
— Tarsi glabrous above. Body unmetallic 64
64(63) Basal margin of elytra strongly arcuate. Labial palp with
  almost cylindrical terminal segment (Fig. 38). (Pl. 6: 9, 10) *Calathus* Bonelli
— Basal margin of elytra not or only slightly arcuate. Labial
  palp with pear-shaped terminal segment (Fig. 37) 65
65(64) Large, about 15 mm long. (Pl. 3: 10) .... *Dolichus* Bonelli
— Smaller, not over 10 mm long. (Pl. 6: 3) .... *Synuchus* Gyllenhal
66(62) Elytron with epipleura "crossed" before apex (Fig. 36) 67
— Elytral epipleura not crossed 69
67(66) Elytra with at least one setiferous dorsal puncture on 3rd
  interval. (Pl. 3: 5 & 6: 5-8) .... *Pterostichus* Bonelli
— Elytron without dorsal punctures 68
68(67) Elytron with 2 extra striae apically outside 8th stria. (Pl. 3: 4) *Abax* Bonelli
— Elytron without supernumerous striae. (Pl. 6: 16-18) .... *Amara* Bonelli
69(66) Elytron without dorsal punctures. (More than 20 mm). (Pl.
  3: 11) .... *Sphodrus* Clairville
— Elytron with at least one dorsal puncture on 3rd interval 70
70(69) Pronotum with two deep, parallel impressions at hind-
  angle. (Upper surface with strong metallic lustre.). ....
  .... *Pterostichus* Bonelli *(metallicus)*
— Pronotum with one or no basal impression laterally 71
71(70) Mentum without tooth (Fig. 22). Pronotum as broad as
  elytra over shoulders. 2nd antennal segment more than
  half length of 3. (Pl. 6: 4) .... *Olisthopus* Dejean

Plate 1

1: Cicindela sylvatica; 2: C. hybrida; 3: C. campestris; 4: Calosoma reticulatum; 5: C. maderae; 6: C. sycophanta; 7: C. investigator; 8: C. inquisitor; 9: Carabus cancellatus; 10: C. granulatus; 11: C. clathratus.

Plate 2

1: Carabus auratus; 2: C. nitens; 3: C. convexus; 4: C. violaceus; 5: C. glabratus; 5: C. coriaceus; 7: C. intricatus; 8: C. problematicus; 9: C. nemoralis; 10: C. arvensis; 11: C. hortensis.

Plate 3

1: Cychrus caraboides; 2: Nebria livida; 3: Broscus cephalotes; 4: Abax ater; 5: Pterostichus burmeisteri; 6: Pt. lepidus; 7: Pt. aterrimus; 8: Pt. niger; 9: Laemostenus terricola; 10: Dolichus halensis; 11: Sphodrus leucophthalmus; 12: Harpalus rufipes; 13: Zabrus tenebrioides.

Plate 4

1: Trachypachus zetterstedti; 2: Omophron limbatum; 3: Leistus rufomarginatus; 4: L. ferrugineus; 5: Nebria brevicollis; 6: Pelophila borealis; 7: Notiophilus biguttatus; 8: Blethisa multipunctata; 9: Diacheila arctica; 10: Elaphrus riparius; 11: E. lapponicus; 12: Loricera pilicornis; 13: Clivina fossor; 14: Dyschirius politus; 15: Miscodera arctica; 16: Patrobus septentrionis; 17: Perileptus areolatus.

Plate 5

1: Aepus marinus; 2: Trechus discus; 3: T. micros; 4: T. rivularis; 5: Asaphidion pallipes; 6: Bembidion litorale; 7: B. pallidipenne; 8: B. varium; 9: B. nitidulum; 10: B. fumigatum; 11: B. bruxellense; 12: B. femoratum; 13: B. lunatum; 14: B. genei; 15: B. quadrimaculatum; 16: B. obscurellum; 17: B. laterale; 18: B. articulatum; 19: B. biguttatum; 20: B. bipunctatum; 21: Tachys bisulcatus, 22: Tachyta nana.

Plate 6

1: Pogonus luridipennis; 2: Stomis pumicatus; 3: Synuchus vivalis; 4: Olisthopus rotundatus; 5: Pterostichus strenuus, 6: Pt. macer; 7: Pt. oblongopunctatus; 8: Pt. nigrita; 9: Calathus fuscipes; 10: C. melanocephalus; 11: Agonum sexpunctatum; 12: A. marginatum; 13: A. dorsale; 14: A. piceum; 15: A. assimile; 16: Amara fulva; 17: A. eurynota; 18: A. aulica.

Plate 7

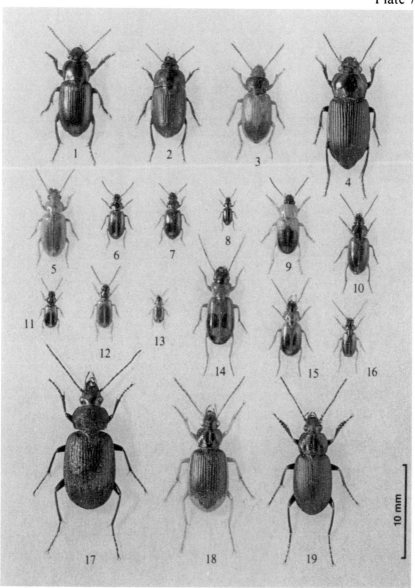

1: Harpalus affinis; 2: H. servus; 3: Diachromus germanus; 4: Anisodactylus binotatus; 5: Dicheirotrichus gustavi; 6: Trichocellus cognatus; 7: Bradycellus harpalinus; 8: B. exiguus; 9: Stenolophus teutonus; 10: S. mixtus; 11: Acupalpus meridianus; 12: A. consputus; 13: Perigona nigriceps; 14: Badister unipustulatus; 15: B. bullatus; 16: B. sodalis; 17: Chlaenius tristis; 18: C. vestitus; 19: C. nigricornis.

Plate 8

1: Licinus depressus; 2: Oodes helopioides; 3: Panagaeus cruxmajor; 4: Odacantha melanura; 5: Masoreus wetterhali; 6: Lebia cruxminor; 7: L. chlorocephala; 8: Demetrias monostigma; 9: D. imperialis; 10: Dromius fenestratus; 11: D. quadrimaculatus; 12: D. sigma; 13: D. melanocephalus; 14: D. linearis; 15: Metabletus truncatellus; 16: Lionychus quadrillum; 17: Microlestes minutulus; 18: Cymindis angularis; 19: Brachinus crepitans.

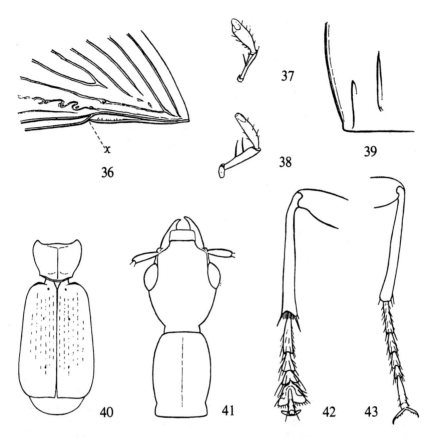

Figs 36-43. 36: apex of left elytron of *Amara,* X = "crossed epipleura"; 37: labial palp of *Synuchus nivalis* (Pz.); 38: same of *Calathus piceus* (Marsh.); 39: hind-angle of pronotum of *Pterostichus lepidus* (Leske); 40: pronotum and hindbody of *Microlestes;* 41: head and pronotum of *Odacantha melanura* (L.); 42: mid-leg of *Demetrias atricapillus* (L.); 43: same of *Dromius longiceps* Dej.

74(73) Claws pectinate. Base of pronotum lobate medially (Fig. 10). Above 5 mm. (Pl. 8: 6, 7)................................ *Lebia* Latreille
  – Claws smooth. Pronotum, Fig. 35. 3-4 mm. (Pl. 8: 16) *Lionychus* Wissmann
75(73) 4th tarsal segment strongly bilobed (Fig. 42). (Pl. 8: 8, 9) *Demetrias* Bonelli
  – 4th tarsal segment with truncate or slightly emarginate apex 76
76(75) Terminal segment of labial palpi dilated and truncate. All elytral intervals finely punctate .... *Cymindis* Latreille *(humeralis)*
  – Terminal segment of labial palpi almost cylindrical. At least not all elytral intervals punctate 77
77(76) Pronotum narrower than head (Fig. 41), both metallic, elytra bicoloured. (Pl. 8: 4) .... *Odacantha* Paykull
  – Pronotum at least as broad as head. Coloration different 78
78(77) Last metatarsal segment equal to first. 3.5-7.0 mm. (only pale or maculate species under 4 mm). (Pl. 8: 10-14) ...... *Dromius* Bonelli
  – Last metatarsal segment shorter than first. Upper surface black (sometimes bronzed). Less than 4 mm 79
79(78) Elytron with apex obliquely truncate and somewhat sinuate (Fig. 34). 3rd antennal segment only with subapical setae. (Pl. 8: 15) .... *Syntomus* Hope
  – Elytral apex transversely truncate (Fig. 40). 3rd antennal segment with sparse pubescence. (Pl. 8: 17) .... *Microlestes* Schmidt-Göbel

SUBFAMILY CICINDELINAE

This has often been regarded as a separate family (Cicindelidae), distinct from the true Carabidae. The main differences are found in the structure of the head (Figs 44, 45): the clypeus and the labrum are very broad, the former wider than the distance between the antennae; the mandibles are armed with several sharp teeth on the internal surface. Unlike the subfamily Carabinae the parameres of the male genitalia are joined by a "basal piece".

The larva lacks cerci on the ninth abdominal segment, and the fifth tergite carries a pair of forwardly directed hooks which support the climbing of the larva in the burrow.

In Europe, except for the extreme south, the subfamily is only represented by the genus *Cicindela*.

Genus *Cicindela* Linnaeus, 1758
(Tiger beetles)

Cicindela Linnaeus, 1758, Syst. Nat. ed. 10: 407.
  Type-species: *Cicindela campestris* Linnaeus, 1758.

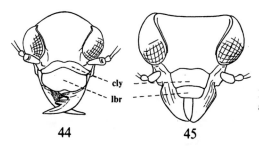

Figs 44, 45. Heads of 44: *Cicindela* and 45: *Elaphrus;* cly = clypeus; lbr = labrum.

Medium sized species (12-19 mm) with head, because of the large semi-globular eyes, at least as wide as pronotum (Fig. 46 & Pl. 1: 1-3). Elytra with pale spots or bands; striae absent (Figs 47-50). Male with 3 dilated pro-tarsal segments, and sixth sternite with a median incision.

The tiger beetles are diurnal predators which occur on sun-exposed ground with sparse vegetation. They are extremely heat-loving insects and active only when the surface temperature is high (Dreisig 1980). During activity they run and fly about with the utmost agility, hunting for insects of all kind. At intolerably high temperatures the beetles show a "stilting"-behaviour and may even bury themselves into the sand. The inactive period is spent in burrows in the ground. Digestion is in *Cicindela* extraintestinal. The Scandinavian species are most numerous in spring and early summer when reproduction takes place. Development lasts two years. The newly emerged imagines are usually fully developed in late summer and normally hibernate in the pupal cells. The larva waits for its prey at the mouth of its vertical burrow in the ground. The biology of *C. hybrida* and *campestris* is treated in detail by Faasch (1968), who also summarizes older literature.

Key to species of *Cicindela*

1   Labrum black, with a median keel. Elytra with irregular, foveate punctures (Fig. 47) .... 1. *sylvatica* Linnaeus
–   Labrum yellow, without keel. Elytra only with dense micropunctures 2
2(1)   Ground colour green; pale elytral markings not confluent (Fig. 50) .... 4. *campestris* Linnaeus
–   Ground colour brownish, sometimes with a metallic hue; pale elytral markings forming a transverse band 3
3(2)   Transverse elytral band more or less angulate (Fig. 49). Frons with a group of erect white setae inside and behind eyes .. 3. *maritima* Dejean
–   Transverse elytral band with a less pronounced bend (Fig. 48). Frons with only 1-3 setae inside hind-margin of eye; in addition, as in *maritima,* with a few setae inside its anterior half .... 2. *hybrida* Linnaeus

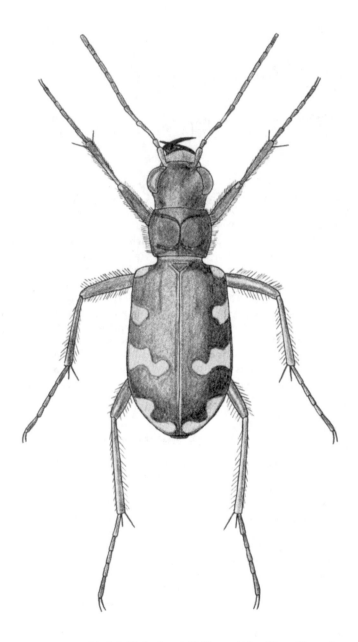

Fig. 46. *Cicindela hybrida* L., length 12-16 mm. (After Victor Hansen).

1. *Cicindela sylvatica* Linnaeus, 1758
   Fig. 47; pl. 1: 1.

Cicindela sylvatica Linnaeus, 1758, Syst. Nat. ed. 10: 407.
Cicindela silvatica auctt.

15-19 mm. Larger and with elytra more stretched than in the other species. Piceous to almost black, usually, at least on forebody, with a bronze hue; underside with a bluish lustre. Of mouth-parts only base of mandibles pale. Elytra (Fig. 47) with a double spot at shoulder, a narrow central band and a rounded subapical spot, yellow. Median parts with coarse, foveate punctures.

Distribution. In Denmark rare but widely distributed in parts of Jutland and in NEZ and B. — Sweden: almost universally distributed and known from all districts, but uncommon and more local in the south as well as in Lapland. — Norway: south of 63°N and in easternmost Finnmark (Fø). — Generally distributed in East Fennoscandia except for the northernmost parts. — Palaearctic region, east to the Amur region; not in the Mediterranean subregion.

Biology. On dry, sandy and sun-exposed ground, mainly in open pine forest with sparse vegetation of *Calluna, Empetrum* and *Cladonia,* but also on heaths and sandy grass areas (e.g. Corynephoretum), and often on recently burned areas. It is strongly heliophilous and heat-preferent. Unlike other species of *Cicindela, sylvatica* often preys upon large ants of the genus *Formica.* It usually flies longer and higher than the other *Cicindela*-species. Mainly in June and July.

2. *Cicindela hybrida* Linnaeus, 1758
   Figs 46, 48, 51; pl. 1: 2.

Cicindela hybrida Linnaeus, 1758, Syst. Nat. ed. 10: 407.

12-16 mm. Bronzy brown with a more or less pronounced greenish hue; underside greenish. Labrum, base of mandibles and usually also the two basal segments of labial palpi testaceous. Elytra (Fig. 48) with a double shoulder spot, one transverse irregular band about middle and another at apex, pale. The central elytral band sometimes a little more angulate than shown in Fig. 48. Penis (Fig. 51) with apex longer and more arcuate than in *maritima;* internal sac with two strong subapical teeth.

Distribution. In Denmark widely distributed and rather common, except in LFM and SZ. — Sweden: rather distributed and not rare in the south-west from Sk. to S.Vrm. — Norway: local in the south-east. — Finland: southern and central parts, more frequent towards the east; also adjacent parts of the USSR. — Entire Europe, Siberia.

Biology. In open, sun-exposed country; especially typical of dry, sandy grass areas; often in dunes and occasionally on the seashore with *C. maritima.* Also on fine gravel, for instance in gravel-pits on moraine. Very thermophilous and in Fennoscandia almost confined to localities with a southern exposure. It appears in early spring, in Denmark already in April, and is active throughout the summer. Young adults are frequently found in the autumn.

3. *Cicindela maritima* Latreille & Dejean, 1822
Figs 49, 52.

Cicindela maritima Latreille & Dejean, 1822, H. N. et Icon. Col. Eur. 1: 52.

12-15 mm. More slender than *hybrida,* notably the pronotum. Otherwise best separated by the elytral pattern. Darker, chocolate-brown, rarely with a greenish hue, appendages coloured as in *hybrida.* Frons less convex anteriorly, inside eyes with a group of erect white setae. Elytra somewhat wider towards apex. Central pale band of elytra more or less angulate at middle (Fig. 49). Meta-tibiae longer and more slender compared to tarsi. Male with longer penis, and internal sac (Fig. 52) differently armed, without apical tooth.

Distribution. Denmark: very rare and local on the coasts; SJ: Rømø; WJ: Fanø, Skallingen; NEJ: Skagen; LFM: Bøtø; B: Rønne, Dueodde. — Sweden: usually very rare, along the coasts of Sk., Hall. and Nb.; also along the rivers in the central and northern parts of the country. No recent records from Sk. and Hall. — Norway: along

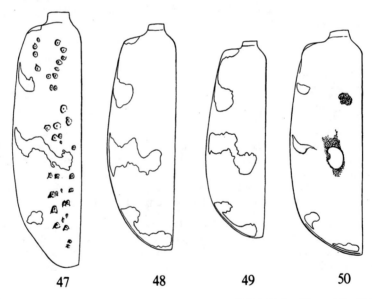

Figs 47-50. Left elytron of *Cicindela.* — 47: *sylvatica* L.; 48: *hybrida* L.; 49: *maritima* Dej.; 50: *campestris* L.

38

rivers in both south and north. — Finland: along the coast and the Bothnian rivers in the north; also Lr: Nuorti. — West coast of Europe and along rivers in the north.

Biology. A stenotopic species, restricted to bare shifting sand near water, mainly on the seashore but also along large rivers; exceptionally on lakeshores. Its habitat is mainly characterised by the grain size and humidity of the sand, not by the salinity. It prefers fine, dry sand. The species is extremely heat-preferent and tolerates temperatures up to almost 50°C (Krogerus, 1932). It is mainly found in May-July.

4. *Cicindela campestris* Linnaeus, 1758
Fig. 50; pl. 1: 3.

Cicindela campestris Linnaeus, 1758, Syst. Nat. ed. 10: 407.

12-16 mm. Separated from the three preceeding species by the green colour and the lack of a central transverse elytral band. Ground colour brilliant green, exceptionally with a bluish hue or, almost black. Labial palpi black, labrum and base of mandibles yellow. Lower surface and abdominal tergites (exposed during flight) bright blue. Elytral pattern (Fig. 50) usually consisting of 5 isolated pale spots (the 2 apical spots sometimes confluent). Male with only extreme apex of mandibles dark. Female almost constantly with a small dark spot in anterior third near the suture.

Distribution. Generally distributed and common in Denmark and Fennoscandia, except in the north. — Entire Palaearctic region.

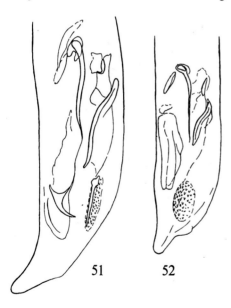

51   52

Figs 51, 52. Apical half of penis of 51: *Cicindela hybrida* L. and 52: *C. maritima* Dej.

Biology. Less thermophilous and more eurytopic than the preceding *Cicindela*-species, occurring in habitats differing much in humidity and vegetation cover. Especially prominent on peaty soil, but also in clayey, sandy and gravelly areas, such as bare ground in *Calluna*-vegetation and sunny woodland paths. It is a typical spring-active species, especially abundant in May, and with only little activity later in the year.

SUBFAMILY TRACHYPACHINAE

An isolated group characterized by the laterally extended meta-coxae (Fig. 53) meeting the margins of elytra (when in repose), and the lack of a ligula in the larva.

Genus *Trachypachus* Motschulsky, 1844

Trachypachus Motschulsky, 1844, Ins. Sibér.: 86.
    Type-species: *Blethisa zetterstedtii* Gyllenhal, 1827.

In general appearance reminding of a large *Bembidion* but with stouter appendages. Frons without furrows. Pronotum with 3 lateral setae (one at front-angle). Elytra with serially arranged punctures but without striae. Hind-wings full. Male with 2 pro-tarsal and one meso-tarsal segments dilated. Parameres symmetrical.
    One palaearctic and 3 nearctic species.

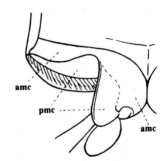

Fig. 53. Metasternal region of *Trachypachus zetterstedtii* (Gyll.), amc = anterior part of meta-coxa, pmc = posterior part of meta-coxa.

5. *Trachypachus zetterstedtii* (Gyllenhal, 1827)
    Figs 53, 54; pl. 4: 1.

Blethisa Zetterstedtii Gyllenhal, 1827, Ins. Suec. 4: 417.

4.5-5 mm. Black and shiny, upper surface with a brass or greenish lustre; legs brown, tibiae palest. Also antennae somewhat pale, with sparse erect setae. Pronotum broadest before middle, latero-basal foveae deep, separated from margin by a strong carina. Rows of elytral punctures evanescent behind middle.

Distribution. Sweden: a few localities from Jmt. north to T. Lpm. — Norway: 3 lo-

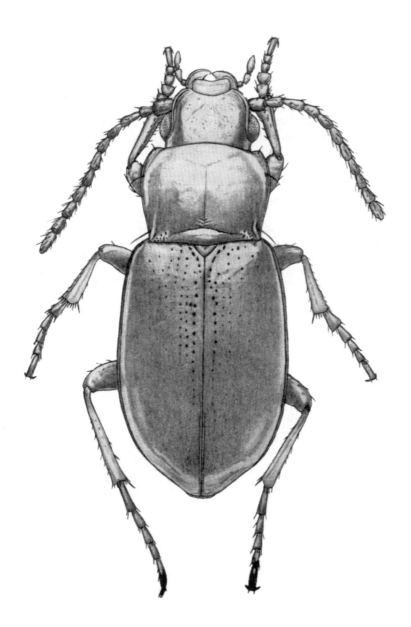

Fig. 54. *Trachypachus zetterstedtii* (Gyll.), length 4.5-5 mm.

calities north of the Polar Circle. — Finland: old records in the inland, in later years only in Lapland. — Northern Eurasia, from Norway to E. Siberia. Very rare and local in Europe, more abundant in Siberia.

Biology. Very little is known about the biology of this rare beetle. It is indigenous to the conifer region. In Sweden (Lu. Lpm.) the species has been found in densely wooded, mixed forest with a rich herb vegetation and a thick layer of litter. The beetles were caught partly by sieving of litter mixed up with decaying remnants of pine and spruce stumps. It is probably diurnal (Lundberg, 1973).

SUBFAMILY OMOPHRONINAE

A small and uniform group, usually considered as a single genus. The habitus reminds of a giant *Haliplus,* which caused earlier authors erroneously to regard the genus as a transition form to the aquatic families of Adephaga.

The body is almost circular (Fig. 56) with pronotum immovably joined to the rest of the body and covering the scutellum. Elytra with supernumerous (in our species 15) striae. A peculiar feature is exposed by the sterna (Fig. 55): the prosternum is enlarged, completely concealing the mesosternum.

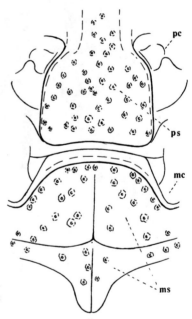

Fig. 55. Sterna of *Omophron,* mc = meso-coxa, ms = meta-sternum, pc = pro-coxa, ps = pro-sternum.

Genus *Omophron* Latreille, 1802

Omophron Latreille, 1802, Hist. Nat. Crust. Ins. 3: 89.
  Type-species: *Carabus limbatus* Fabricius, 1777.

Ground colour of body and appendages pale, elytra with dark pattern. Wings full.
Male with two pro-tarsal and one meso-tarsal segments dilated. The penis has an
"open" (non-sclerotized) dorsum; the parameres are non-setose, pointed, similar but
not symmetrical.

6. *Omophron limbatum* (Fabricius, 1777)
  Figs 55, 56; pl. 4: 2.

Carabus limbatus Fabricius, 1777, Gen. Ins.: 240.

5-6.5 mm. Pale yellowish brown, most of head, centre of pronotum, and 3 irregular
transverse bands on elytra, joined along the suture, metallic green.

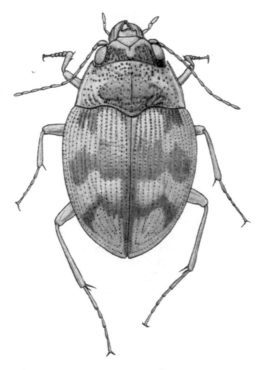

Fig. 56. *Omophron limbatum* (F.), length 5-6.5 mm.

Distribution. Denmark: in Jutland only a few old records from SJ and EJ; very local on the islands, but more abundant on B. — Sweden: rather distributed in the south-west, from Sk. to Vg.; also Ög. — Absent from Norway and East Fennoscandia. — Most of Europe, W.Asia, Siberia.

Biology. A strictly riparian species, living on shores of lakes, ponds and streams; also on sea-slopes with out-welling freshwater. Always found on bare sand or sand mixed-up with clay. It is a predatory beetle, hunting and flying at night. During day-time it stays in burrows in the sand and is easily forced out by splashing the surface with water. Mainly in May-August. Fully grown larvae occur from spring to late summer. This caused Larsson (1939) to believe that the development lasts two years. This can not been confirmed. The digestion is not extra-intestinal as in most other Carabidae.

SUBFAMILY CARABINAE

Since no characters have been found for defining the "Harpalinae" as a separate sub-family, all the following genera, except *Brachinus,* are included in the Carabinae. This includes the overwhelming majority of genera. For diagnostic characters, see diag-noses of other subfamilies.

Tribe Carabini

This group includes some of the largest and most well-known ground-beetles. They deviate from most other Carabids by the elytral sculpture, which either consists of su-pernumerous (more than 15) regular striae, or of an irregular system, often including tubercles, ridges, foveae, etc.; or the elytra are virtually smooth.

Genus *Calosoma* Weber, 1801

Calosoma Weber, 1801, Obs. Ent. 1: 20.
   Type-species: *Carabus sycophanta* Linnaeus, 1758.

Our species are easily separated from *Carabus,* besides by the general habitus (Fig. 57), on the wrinkled mandibles and the very short second antennal segment (Fig. 58). The pronotum is much smaller than in *Carabus,* both narrower and shorter as compared with the elytra. Wings always full and functionary. Upper surface with a more or less pronounced metallic reflection. The elytral sculpture is much varying, with 3 rows of small foveae, and between these either with numerous regular striae, or with irregular and more or less confluent structures.

   In warmer regions of both the Old and New World the genus contains very *Carabus*-like species with reduced wings. *Calosoma* is particularly numerous in species in the Nearctic region. Both adults and larvae are voracious predators, notably on caterpil-lars, and some species *(inquisitor, sycophanta)* are arboreal rather than terricolous and climb trees. Any morphological adaptations to this mode of life can not be detected in these species. Adults occur in early summer and then aestivate.

Of the 6 species in our area only two *(inquisitor* and *reticulatum)* have established permanent populations.

Fig. 57. *Calosoma inquisitor* (L.), length 16-22 mm. (After Victor Hansen).

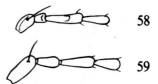

Figs 58, 59. Antennal base of 58: *Calosoma inquisitor* (L.) and 59: *Carabus violaceus* L.

58

59

Key to species of *Calosoma*

1   Elytra with well-marked punctate striae over entire surface; transverse lines of intervals weak (Fig. 57) 2
–   Elytral striae obliterated, at least towards apex, and intruded upon by punctate, granulate or transverse sculpture 3
2(1) Pronotum and elytra of different colour. Elytral intervals with extremely fine transverse lines. Large species: over 24 mm .... 7. *sycophanta* (Linnaeus)
–   Pronotum and elytra of same colour. Transverse elytral lines coarse, at least laterally. Smaller species, less than 22 mm .... 8. *inquisitor* (Linnaeus)
3(1) Elytra with 3 rows of metallic foveae 4
–   Elytral foveae not differently coloured, difficult to discern because of the coarse background sculpture. Sides of pronotum rounded throughout (Fig. 65) .... 12. *reticulatum* (Fabricius)
4(3) Sides of pronotum straight or slightly rounded posteriorly (Fig. 62). Elytra between rows of foveae with 3 rows of regular scales .... 9. *maderae auropunctatum* (Herbst)
–   Sides of pronotum more or less sinuate before hind-angles (Figs 63, 64). Scales of elytra granulate, arranged in about 5 irregular rows between the serial foveae 5
5(4) Hind-angles of pronotum angulate, prominent (Fig. 63) . 10. *denticolle* Gebler
–   Hind-angles of pronotum obtuse (Fig. 64) .... 11. *investigator* (Illiger)

7. *Calosoma sycophanta* (Linnaeus, 1758)
Fig. 61; pl. 1: 6.

Carabus sycophanta Linnaeus, 1758, Syst. Nat. ed. 10: 414.

24-30 mm. Black with a bluish lustre, especially on pronotum. Elytra brilliantly green, often with a reddish hue. Pronotum (Fig. 61) with a complete raised margin, hind-angles projecting more directly backwards than in *inquisitor.* Male with 3 dilated protarsal segments.

46

Distribution. Denmark: records of single specimens from the islands, SJ and EJ, but obviously becoming rarer in this century; only one capture after 1950 (NWZ: Skellebjerg, 26.VII.1958). — Sweden: very rare and single, mainly in the south, northernmost capture in Dlr. — Not in Norway. — In Finland one accidental introduction (N: Grankulla 1982). — Main part of C. and S.Europe, N.Africa, W.Asia. Introduced in N.America for regulation of destructive moths and now well established in eastern USA.

Biology. An excellent flyer and in our area only occurring as an accidental visitor, strayed from the south. In C.Europe it is an inhabitant of both deciduous and coniferous forests. Preferably on, but not so strictly associated with, oak as is *C. inquisitor.* Rather unfastidious, preying upon larvae of a number of different moths (e.g. Lymantriidae, Thaumatopoeidae). Both adults and larvae climb trees in search of caterpillars.

Note. Nearest breeding localities are in N.Germany (Horion, 1941). Numerous subfossil remains from the postglacial warm period have been discovered in Sweden (Lindroth, 1949), and a permanent population was no doubt then present.

8. *Calosoma inquisitor* (Linnaeus, 1758)
   Figs 57, 58, 60; pl. 1: 8.

Carabus inquisitor Linnaeus, 1758, Syst. Nat. ed. 10: 414.

16-22 mm. Black, greenish brass underneath, upper surface more or less bronze, often with a reddish or greenish hue; sides of elytra with a stronger, usually pure green lustre. Male with 4 dilated pro-tarsal segments. Meso-tibiae somewhat arcuate.

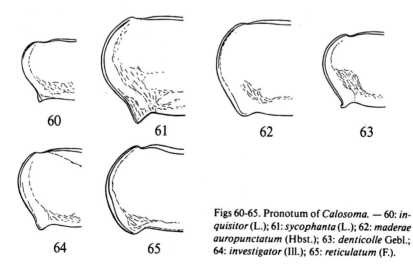

60

61

62

63

64

65

Figs 60-65. Pronotum of *Calosoma.* — 60: *inquisitor* (L.); 61: *sycophanta* (L.); 62: *maderae auropunctatum* (Hbst.); 63: *denticolle* Gebl.; 64: *investigator* (Ill.); 65: *reticulatum* (F.).

Distribution. Very local in Denmark, earlier frequently recorded but only 5 records after 1950. — Sweden: rather distributed from Sk. to S.Dlr. — In Norway recorded from a few southern districts. — In Finland permanently only in Ab: Turku, Ruissalo; one stray individual N: Tvärminne. Not in adjacent parts of the USSR. — Almost entire Europe, N.Africa, W.Asia, E.Siberia, Japan.

Biology. In open deciduous forest, notably of oak. Only the adults are arboreal and hunt during daytime. The larvae are ground-dwellers. A more oligophagous species than *C. sycophanta,* the prey consisting of larvae and pupae, mostly of Geometridae (e.g. *Operophtera brumata* L.) and Tortricidae (e.g. *Tortrix viridiana* L.). The species is local and periodical. It may suddenly appear in numbers after years of absence, and this phenomenon has probably a correlation with outbreaks of lepidopterous larvae. Predominantly in June.

9. *Calosoma maderae* (Fabr.) ssp. *auropunctatum* (Herbst, 1784)
  Fig. 62; pl. 1: 5.

Carabus Maderae Fabricius, 1775, Syst. Ent.: 237.
Carabus auropunctatum Herbst, 1784, Arch. Insectengesch. 5: 131.
Carabus sericeus Fabricius, 1792, Ent. Syst. 1: 147.

20-30 mm. Narrower, with longer and more parallel-sided elytra than in any other species. Black, upper surface with a faint bronze, often somewhat greenish, hue. Elytra with 3 rows of golden or greenish foveae. Distinguished from the 3 following species, partly by the form of the pronotum (Fig. 62), partly by the elytral structure which consists of scales arranged in regular rows between the rows of foveae. Male with 3 dilated pro-tarsal segments; meso- and meta-tibiae strongly arcuate.

Distribution. Apparently only a casual visitor from the south. — Denmark: the few records are all from coastal localities. Only one record after 1950, SJ: Rømø, 21.V.1961. The records from EJ, NEJ, LFM and NEZ are from before 1900. — Sweden: 5 records from Sk., Hall. and Öl.; most records are old, the most recent is Öl., Bredsätra, 1937. — Norway: one old specimen, from near Oslo. — Not in E.Fennoscandia.

Biology. On open, sandy areas, in Scandinavia almost restricted to coastal commons. In C.Europe also on inland localities, frequently occurring in numbers in cultivated fields. It is an entirely terricolous predator of larvae of moths (e.g. Noctuidae) and other insects. Breeding takes place in May and June. Unlike the preceding species the newly emerged adults of *C. maderae* are active in late summer before the hibernation.

10. *Calosoma denticolle* Gebler, 1833
  Fig. 63.

Calosoma denticolle Gebler, 1833, Bull. Soc. Nat. Mosc. 6: 274.

19-26 mm. At once recognized by the acute, dentiform hind-angles of pronotum (Fig.

48

63). The elytra are longer and more parallel-sided than in *investigator.* Also the serial foveae are smaller and denser, and the sternites have a sparser provision of setae. Male without dilated protarsal segments.

Distribution. An accidental visitor to our area, Finland: N, Tvärminne, Brännskär, 30.VIII.1935 (R.Storå). — The distribution extends over the southern parts of Russia and Siberia.

11. *Calosoma investigator* (Illiger, 1798)
Fig. 64; pl. 1: 7.

Carabus investigator Illiger, 1798, Verz. Käf. Preuss.: 142.

16-23 mm. Smaller and shorter than *auropunctatum,* otherwise similar. Black, rather shiny, upper surface faintly bronze, never greenish. Sides of pronotum (Fig. 64) with a shallow but evident sinuation posteriorly. The sculpture of elytra dense, consisting of granulae without clear formation of scales. Male with 3 dilated pro-tarsal segments, meso-tibiae slightly arcuate.

Distribution. An accidental visitor. Sweden: Öland, one old specimen without exact locality (A. Mortonson, Mus. Göteborg). — Palaearctic, east to Amur. Nearest localities are in NE.Germany.

12. *Calosoma reticulatum* (Fabricius, 1787)
Fig. 65; pl. 1: 4.

Carabus reticulatus Fabricius, 1787, Mant. Ins. 1: 197.

20-26 mm. The pronotum is very broad when compared with the elytra, which gives the species a *Carabus*-like appearance. Black, upper surface brilliantly emerald green. Old specimens are dark, often quite black, but the green colour remains on elytral epipleura and underside of pronotum. Sides of pronotum (Fig. 65) strongly convex. The 3 series of elytral punctures are small, not differently coloured and difficult to discern. The ground sculpture consists of very coarse granulae, in part fused into irregular transverse wrinkles. Male with 3 dilated pro-tarsal segments. Meso-tibiae only slightly arcuate.

Distribution. Denmark: only 7 records (all earlier than 1900) from SJ (5 records) and EJ (Vejle, Kolding). — Sweden: besides 2 specimens from Boh., Foss, Torreby (Notini) and one old specimen from Sk., only known from Öland, where it has a constant, though small, population on the open steppe-like "alvar" in the southern part of the island. — Not known from Norway or Finland. — Very restricted: from N.Germany to Poland, south into C.Europe.

Biology. A xerophilous species, occurring on dry, sandy localities such as heaths and open pine forest, often near the coast. On the "alvar" steppe of Öl. (see above). It is a ground-dweller which rarely climbs the vegetation. Mainly in June.

Genus *Carabus* Linnaeus, 1758

Carabus Linnaeus, 1758, Syst. Nat. ed. 10: 415.
  Type-species: *Carabus granulatus* Linnaeus, 1758.

Body rather slender with pronotum not much narrower than elytra. Appendages slender. Elytral sculpture never regularly striate but consisting of carinae, tubercles and foveae, often with interlying very dense striae, or is quite irregular or almost smooth. Wings rudimentary, except full and functionary in single individuals of *clathratus* and (in C.Europe) *granulatus*. Male with 3 or 4 strongly dilated pro-tarsal segments.

Carabus was used by Linnaeus and Fabricius to include nearly all of their Carabid species but was constricted to its present content after 1800. Attempts to split it into several genera is seldom accepted.

The infraspecific variation, notably of the elytral sculpture, is more pronounced in *Carabus* than in any other genus and this has caused the creation of an almost unsurveyable abundance of names, of subspecific or lower range. They have been summarized in Breuning's monograph (1932-37), but his application of a strictly quaternary, and even quintenary, nomenclature is not in accordance with the *International Code of Zoological Nomenclature* (1961). It has not been considered necessary to use here the numerous subgeneric names of *Carabus,* but the species are arranged in the same order as in Breuning's monograph. With the exception of *cancellatus* and *problematicus,* the 16 Fennoscandian species are morphologically rather stable.

The species are basically predators without pronounced specialisation, preying upon insects of almost every kind, and upon earthworms, slugs, etc. However, they do also feed on carrion and sometimes even on vegetable matters. Digestion is extra-intestinal. A number of *Carabus*-species are known to live as adults for several years and to reproduce in several successive breeding seasons. The phenology of C.European species was described in detail by Hůrka (1973).

Key to species of *Carabus*

1	Elytra each with 2-4 continuous (rarely partly interrupted) elevated carinae (Figs 67-69)	2
–	Elytra without or with more numerous, less pronounced ridges	7
2(1)	Each interval between the carinae with a single row of tubercles or foveae (Figs 67, 68)	3
–	Intervals without such longitudinally arranged sculpture (Fig. 69)	6
3(2)	Intervals foveate (Fig. 68)	17. *clathratus* Linnaeus
–	Intervals tuberculate (Fig. 67)	4
4(3)	First antennal segment rufous. Apical setae of antennal segments 3 and 4 of equal density	18. *cancellatus* Illiger
–	Antennae entirely black. Apical setae of antennal segment 4 denser than of segment 3	5

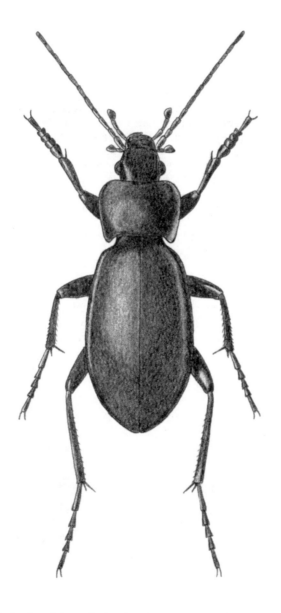

Fig. 66. *Carabus violaceus* L., length 20-30 mm. (After Victor Hansen).

5(4)   The two strongest elytral carinae (notably towards apex) somewhat irregular and with sparse punctures; a weak but continuous carina present between suture and innermost row of tubercles (Fig. 67) . . . . 15. *granulatus* Linnaeus

–   The two main elytral carinae stronger and quite smooth; a sutural carina absent or only suggested anteriorly . . 16. *menetriesi* Hummel

6(2)   Elytral carinae black, unmetallic. Appendages black . . . 24. *nitens* Linnaeus

–   Elytral carinae metallic as is the background. Basal antennal segments, femora and tibiae rufous . . . . 14. *auratus* Linnaeus

7(1)   First antennal segment without setae. Sculpture of elytra irregular, reticulate. Entirely black species, 32-40 mm 28. *coriaceus* Linnaeus

–   First antennal segment with setiferous punctures. Elytral sculpture not reticulate. Smaller species, rarely over 30 mm 8

8(7)   Elytra with clearly longitudinally arranged sculpture, consisting of fine ridges and rows of small tubercles and/or foveae (Fig. 70) . . 9

–   Elytral sculpture weak and irregular, rarely with 3-4 faint elevated lines caused by confluence of granulae (Fig. 66) 15

9(8)   Each elytron with 3 rows of golden or greenish foveae on black background . . . . 21. *hortensis* Linnaeus

–   Foveae small and not differently coloured, or are absent 10

10(9) Penultimate segment of maxillary palpi (the larger pair) shorter than the terminal segment. Elytra each with more than 20, usually weak ridges, and 3 rows of small punctures . . . . 25. *convexus* Fabricius

–   Penultimate segment of maxillary palpi at least as long as

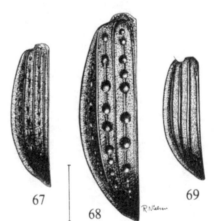

Figs 67-69. Left elytron of *Carabus*. —
67: *granulatus* L.; 68: *clathratus* L.;
69: *nitens* L.

terminal segment. Elytral ridges fewer, more or less un-
equal, often interrupted 11

11(10) Forebody very narrow, pronotum not wider than long. Distance between eyes equal to distance from centre of eye to
base of labrum.............................. 26. *intricatus* Linnaeus
- Pronotum wider than long. Distance between eyes much
longer than from centre of eye to labrum 12

12(11) Ridges between the three elytral rows of punctures well developed and smooth .... 13. *monilis* Fabricius
- Elytra with weak, more or less irregular ridges on the intervals between the rows of foveae, or are otherwise sculptured 13

13(12) Elytral intervals without evident ridges, sculpture irregular
and scale-like .... 20. *nemoralis* Müller
- Each elytral interval at least anteriorly with three ridges
(Fig. 70) 14

14(13) Penultimate segment of labial palpi (the smaller pair) with
several setae. Pronotum with greatest width in anterior
third, sides elevated basally .... 23. *problematicus* Herbst
- Penultimate segment of labial palpi bisetose. Pronotum
widest near middle, sides not elevated .... 14. *arvensis* Herbst

15(8) Penultimate segment of labial palpi with several setae.
Forebody clearly metallic along side-margins...... 27. *violaceus* Linnaeus
- Penultimate segment of labial palpi bisetose. Upper surface
unicolorous black, often with more or less distinct steel-
blue lustre .... 22. *glabratus* Paykull

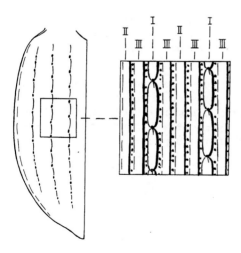

Fig. 70: Left elytron of *problematicus*
Hbst. — I = primary, II = secondary,
III = tertiary carinae (according to
Breuning).

53

13. *Carabus monilis* Fabricius, 1792

Carabus monilis Fabricius, 1792, Ent. Syst. 1: 126.

22-26 mm. In general habitus similar to *violaceus,* but pronotum broader and more like that of *problematicus;* legs shorter. Black, usually with a coppery or varying metallic lustre, sides of pronotum and elytra more brilliantly coloured. Intervals between row of elytral punctures ("II-III" in Fig. 70) with three smooth ridges, the central of which may be more strongly developed.

Distribution. In our area only found in SE. Norway, in the town of Fredrikstad; no doubt originally introduced but possibly now naturalized. — European species, north to N.Germany.

Biology. In C.Europe this is a eurytopic species associated with open, cultivated land; also along river banks and to a lesser extent in woodland. An autumn breeder.

14. *Carabus arvensis* Herbst, 1784
    Pl. 2: 10.

Carabus arvensis Herbst, 1784, Arch. Insectengesch. 5: 132.
Carabus arcensis auctt.

15-20 mm. Black with a various degree of metallic lustre: violaceous, greenish, coppery, etc. Penultimate segment of labial palpi bisetose. Pronotum conspicuously flat, sides little elevated; its greatest width occurs about middle. The elytral sculpture is most similar to *problematicus* but the interrupted primary ridges are stronger. Male with antennal segments 6-8 slightly concave underneath.

Distribution. Denmark: Jutland, NEZ and NWZ, but rare and very local. Also SZ: Sorø; not in F, LFM and B. — Sweden: Sk. to Hls., local and rare. — Norway: only in some southern districts. — Finland: rare in the southern parts, north to Tb; also in the Svir district of the USSR. — Almost entire Europe except the south; Siberia east to the Pacific Ocean.

Biology. A xerophilous species, notably on gravel or sand in open, dry country. Especially characteristic for *Calluna* heaths with scattered pine growth and forest clearings on burned areas. It is mainly diurnal. The adults have their peak of activity in May-June when they reproduce. The newly emerged beetles occur in late August and September. They hibernate under bark of tree-stumps, in moss, etc.

15. *Carabus granulatus* Linnaeus, 1758
    Fig. 67; pl. 1: 10.

Carabus granulatus Linnaeus, 1758, Syst. Nat. ed. 10: 413.

16-23 mm. A slender species with a narrow pronotum. Upper surface black, almost constantly with a brass or greenish reflection. Antennae and legs usually entirely black (see note below). Among species with carinate elytra it is characterized by the strongly elevated and posteriorly sinuate sides of pronotum. Elytral intervals with strong tubercles (Fig. 67). Apical setae of antennal segment 4 denser than of segment 3. Female with elytral margin deeply sinuate before apex.

Distribution. Denmark: generally distributed and very common. — Sweden: Sk. to Ång., generally distributed in the south. — Norway: mainly along the south coast up to the Bergen-area (HO). — East Fennoscandia: rather frequent in the southern half, north to the Oulu-area in Finland and to Paanajärvi on the Oulanka river. — Entire Europe, Siberia, to the Pacific Ocean. Introduced in N.America.

Biology. The species is hygrophilous, occurring in wet meadows and in open, deciduous forests on moist, clayish soil; often in stands of alder *(Alnus)* along river banks and lake shores; also on arable land. It is nocturnal and a spring breeder with summer larvae. Reproduction takes place mainly in May-June, and the new generation of beetles occurs from late July to September. *C. granulatus* spends the daytime and the winter in tree stumps and felled timber.

Notes. Specimens with full hind-wings and power of flight are known from C.Europe but not from our area.
The pale-legged form is in Denmark especially abundant in the western and northern parts of Jutland. Only a few pale-legged specimens are known from the islands (Bangsholt, 1983: 85).

16. *Carabus menetriesi* Hummel, 1827

Carabus menetriesi Hummel, 1827, Essais Ent. 6: 3.

16-22 mm. Very similar to *granulatus* but deviating in the following respects: pronotum broader, more convex and with a more shallow sinuation of the sides. Punctuation of forebody coarser and less dense. Elytra more convex with stronger tubercles; the short sutural ridge absent or present as a rudiment near the base. The subapical sinuation of the female elytral sides is little pronounced.

Distribution. USSR: Vib, many localities, some quite near the Finnish border; Kr, several localities near river Svir. — Locally in Central Europe from Bavaria and Austria to the East Baltic area.

Biology. In forested swamps, pronouncedly hygrophilous and reported to live as *clathratus*.

17. *Carabus clathratus* Linnaeus, 1761
    Fig. 68; pl. 1: 11.

Carabus clathratus Linnaeus, 1761, Fauna Suec. ed. 2: 218.

22-28 mm. The elytral sculpture (Fig. 68) is unique. Black, including appendages, upper surface almost constantly with a greenish brass reflection, the intercostal elytral foveae golden or coppery. Pronotum very broad with deep basal foveae and protruding hind-angles. Normally with quite reduced hind-wings, but certain populations, especially around the northern parts of the Bothnian Bay, contain long-winged individuals, capable of flying.

Distribution. In Denmark rare and very local, recorded with decreasing frequency. — Sweden: rare and local, occurring in two mains areas: locally in the south from Sk. to Upl., further Vb. to Lu.Lpm. Between these areas only one locality in Hls. — Norway: very local in the south. — Finland: north to LkW; also adjacent parts of the USSR. — Most of Europe, except western and southern parts, but present in Scotland and Ireland. N.Asia to the Pacific Ocean.

Biology. A very hygrophilous species, occurring on muddy lake shores, in swamps and peat bogs where the vegetation is luxuriant; rarely on moderately wet peaty soil. In Denmark also in salt meadows near the sea. It may enter down into the water and clamber about among the water plants to hunt for snails, crustaceans, leeches, insect larvae and other prey, and is able to stay submerged for more than 15 minutes (Sturani 1962). A bubble of air is stored under the elytra for the respiration. Predominantly nocturnal, but also occasionally running about in sunny weather. The species is a spring breeder, is seen most numerous in May-June; only little activity in the autumn.

18. *Carabus cancellatus* Illiger, 1798
   Pl. 1: 9.

Carabus cancellatus Illiger, 1798, Verz. Käf. Preuss.: 154.

22-27 mm. At once recognized by the rufous first antennal segment; otherwise similar to *granulatus*. Stout and convex. Upper surface with bronze or coppery lustre, most pronounced on the pronotum. Apical setae of antennal segments 3 and 4 equal. Pronotum more densely punctate than in *granulatus* and with backwardly produced hind-angles; latero-basal foveae shallow. Elytra almost smooth between ridges and rows of tubercles. Female with margin of elytra angulately incised before apex.

Distribution. Denmark: rare and with scattered distribution; has decreased recently; only 14 captures since 1950. — Sweden: recorded with decreasing frequency and now usually rare. In the SW, north to Vrm., and east to Sm. and Öl. — Norway: in the south-eastern districts, also one loc. in SFi. — East Fennoscandia: in the central and south-eastern parts of Finland, north to Kuopio, common in the east; also Vib and Kr in the USSR. — Almost all of Europe, absent from the British Isles. Siberia east to Lena.

Biology. In open land, preferably on clayey soil with some moisture; often on arable land, but also on sandy grass areas and in forest clearings. It is predominantly nocturnal but shows additional diurnal activity in the reproductive period. Breeding takes place in spring, mainly in May-June; newly emerged beetles occur in July-August.

Note. In C.Europe this species has been divided into numerous subspecies and subordinate forms. It is also more variable in Denmark than in Scandinavia (see Bangsholt, 1983: 86). Individuals with rufous or brownish femora seem to appear almost everywhere as an aberration, and are in Denmark more abundant in Jutland than on the islands. In populations from Finland and adjacent parts of the USSR (which have immigrated from the south-east), this character is constant and also correlated with stronger elytral tubercles.

19. *Carabus auratus* Linnaeus, 1761
Pl. 2: 1.

Carabus auratus Linnaeus, 1761, Fauna Suec. ed. 2: 219.

20-27 mm. Easily recognized by the uniformly metallic green upper surface, except that the elytral margins are golden. The 4 basal segments of antennae, femora and tibiae rufous. The elytra carry 4 broad continuous carinae with virtually smooth intervals. Female elytra as in *cancellatus*.

Distribution. This species is rapidly expanding westward on the European continent. The few finds in our area are no doubt due to accidental introduction. — Denmark, Sweden and Norway: a few probably introduced specimens; most recently found on Langeland (Martin, 1983). Not in East Fennoscandia.

Biology. In C.Europe this species prefers warm, open areas, mainly on heavy clay soil, often in cultivated fields. It is a spring breeder with pronounced diurnal activity.

20. *Carabus nemoralis* Müller, 1764
Pl. 2: 9.

Carabus nemoralis Müller, 1764, Fauna Ins. Fridr.: 21.

22-26 mm. A stout, rather convex and short-legged species without elytral ridges. Bronze to brass green, the female more dull, sides of pronotum and elytra more or less violaceous. Each elytron with 3 rows of small foveae which sometimes are more metallic than the background. Sculpture otherwise irregular, scale-like. Male more shiny, less convex. Antennal segments 6-8 slightly convex below.

Distribution. Widely distributed and common in Denmark. — In Sweden very distributed and common in the south, north to the Bothnian Bay in Hls. and Med.; after 1950 also in Ång. and Nb. — Norway: along the coast to 64°N. — East Fennoscandia: common in the southern half of Finland, and in Vib and the Svir district of the USSR. — Almost entire Europe, except in the extreme south and east. Introduced in N.America.

Biology. A very eurytopic species, occurring on all kinds of moderately dry soil rich in humus: in light forests, parks and gardens, even in the middle of towns, as well as in open country, especially on arable land. The species is clearly favoured by agriculture. It is nocturnal and a spring breeder, reproducing in April-June. Newly emerged adults occur in August-September. The species in biennial in northern Fennoscandia.

Note. The most abundant *Carabus* of Scandinavia. It has obviously expanded considerably during the last century and was unknown to Linnaeus. It has probably intruded upon the habitat of *C. hortensis*.

21. *Carabus hortensis* Linnaeus, 1758
Pl. 2: 11.

Carabus hortensis Linnaeus, 1758, Syst. Nat. ed. 10: 414.

22-28 mm. More slender than *nemoralis* and with a narrower pronotum and longer legs; at once recognized by the elytral foveae. Black, elytra slightly metallic (usually violaceous). Along each elytral margin 3 rows of deep golden (rarely greenish) foveae and between these very fine, regular longitudinal striae. Male more parallel-sided than female. Antennal segments 5-9 sinuate below.

Distribution. Denmark: widely distributed and rather common except in the western and northern parts of Jutland. — On the Scandinavian peninsula and in E.Fennoscandia distributed to about 64°N; common in S. & C.Finland. — A strictly European species, but lacking in entire western Europe.

Biology. A typical forest species. In Scandinavia almost restricted to deciduous and mixed forests on humus-rich, rather dry soil; in C.Europe more eurytopic, occurring also in coniferous stands. It is a nocturnal autumn breeder which reproduces mainly in August-September. Not only the larvae, but also a considerable portion of the adults, hibernate. In Denmark, Schjøtz-Christensen (1968) found that about 25% of a *C.hortensis*-population entered upon a second breeding period following hibernation. The species is biennial in northern Fennoscandia.

22. *Carabus glabratus* Paykull, 1790
Pl. 2: 5.

Carabus glabratus Paykull, 1790, Mon. Car. Suec.: 14.

22-30 mm. Very convex and with a short pronotum. Unicolorous coal black, sometimes with a steel-blue lustre, more shiny than normal *violaceus*. Can by separated from that species by the bisetose penultimate segment of labial palpi. Pronotum with confluent, wrinkled punctuation inside hind-angles. The elytral sculpture consists of small, flat granulae, exceptionally joined to form three very obsolete raised lines on each elytron. Male more parallel-sided than female, with apex of elytra less acuminate. Antennal segments 6-8 sinuate below.

Distribution. Denmark: abundant around Silkeborg and Århus (both in EJ) and in NEZ; otherwise very local, apparently absent from many parts of the country. — Sweden, Norway and Finland: generally distributed and common. — Entire Europe, except in the south-east, south and south-west.

Biology. In the lowland exclusively in forests, notably dark moist spruce-mixed forests rich in mosses, but also in beech forests. In mountainous regions where the species is more abundant, it is also an inhabitant of open country, in the alpine region often on dry soil with sparse vegetation. In the south, *C.glabratus* is a nocturnal autumn breeder having winter larvae. In Poland, Grüm (1979) found that some adults may hibernate and reproduce for a second time in early summer. In northern climates the species is predominantly diurnal and has a biennial life cycle with reproduction in spring (Houston, 1981).

23. *Carabus problematicus* Herbst, 1786
   Fig. 70; pl. 2: 8.

Carabus problematicus Herbst, 1786, Arch. Insectengesch. 6: 177.
Carabus catenulatus auctt.; *nec* Scopoli.

18-30 mm. A convex species with uniform elytra and slender legs. Black, with a more or less distinct, usually violaceous (sometimes greenish or coppery) lustre along the margins of pronotum and elytra. Pronotum with greatest width in anterior half, sides reflexed and elevated basally. Characterized by the rough, complex, irregular elytral sculpture (Fig. 70): the "primary" ridges are interrupted by small setiferous punctures, on the intervals weaker; entire "secondary" and "tertiary" ridges obsolescent towards apex.

   In this species a division into subspecies is more appropriate than in any other Scandinavian *Carabus* (Strand, 1935).

23a. *Carabus problematicus gallicus* Géhin, 1885

Carabus problematicus var. *gallicus* Géhin, 1885, Cat. Col. Carab.: 15.
Carabus catenulatus var. *scandinavicus* Born, 1926, Norsk Ent. Tidsskr. 2: 63.

23-30 mm. More slender, with narrower pronotum and longer legs than in the two following subspecies. The colour of the margins always markedly violaceous. Elytral sculpture more regular.

   Distribution. Denmark: only in Jutland where widely distributed and rather common, but less so in the east and south of the peninsula. — Sweden: southern and central parts, north to Dlr., usually rare. — Norway: along the coast north to 65°N. — Not in Finland. — With a southern distribution in Europe.

23b. *Carabus problematicus wockei* Born, 1898

Carabus catenulatus Wockei Born, 1898, Soc. Ent. 13: 74.

19-24 mm. Body stouter. Metallic lustre more bluish, often quite weak (rufinistic individuals not uncommon). Elytral sculpture usually more irregular. Transitional forms (probably hybrids with *gallicus*) occur in SW. Norway.

Distribution. In the southern half of the Scandinavian mountain range. Sweden: Dlr., Hrj. and Jmt. — Norway: north to the Trondheim area. — Absent from East Fennoscandia, but a doubtful form exists on the island Hogland in the Gulf of Finland (*relictus* Hellén, 1934).

23c. *Carabus problematicus strandi* Born, 1926

Carabus catenulatus strandi Born, 1926, Norsk Ent. Tidsskr. 2: 65.

18-22 mm. More slender than *wockei* and with more parallel-sided elytra; their sculpture irregular and more shiny.

Distribution. Exclusively northern. Norway: north of the Polar Circle, from Bodö to the extreme north. — Sweden: T. Lpm., Karesuando. — Finland: recorded once in Le: Enontekiö, Urttaspahta (near Halti) and once in Li: Utsjoki. — USSR: Kola Peninsula. Not found east of the White Sea.

Biology. A heat-preferent species, in Scandinavia favouring open, dry habitats on gravelly or sandy soil, especially moraine. Mostly on heather *(Calluna),* but also in open pine forest; in the alpine region on dry heaths. In Denmark (Jutland) it is a rather common inhabitant of light deciduous forests, notably of oak, and it is considered a true forest species in C.Europe. Strictly nocturnal. In southern Scandinavia the species is an autumn breeder. In Holland, Drift (1959) found *C.problematicus* to breed in August-October. The larvae hibernated, and the new generation of beetles emerged in the following spring. They entered upon an aestivation dormancy lasting until the onset of reproduction. A number of adults hibernated and bred for a second time. In northern climates the life cycle may be biennial, with adults emerging in late summer or autumn and breeding in the following autumn (Houston, 1981).

24. *Carabus nitens* Linnaeus, 1758
Fig. 69; pl. 2: 2.

Carabus nitens Linnaeus, 1758; Syst. Nat. ed. 10: 414.

13-18 mm. Generally regarded as the prettiest of our ground beetles. Black; elytra, except the pure black ridges, metallic green; pronotum, often head and elytral margin, golden or coppery. Rare aberrations may have the pronotum green, the elytra golden red, or the entire upper surface almost unmetallic black (the latter form is particularly found in N. Finland). Antennae very short, segments 2-3 flattened. Apex of pro-tibia hooked. Elytral ridges often interrupted, notably apically (ab. *fennicus* Géhin), intervals with a faint transverse sculpture.

Distribution. Denmark: rare and very local, absent from several districts on the islands; has decreased strongly recently- — Sweden: rather distributed but not common, in some areas strongly decreasing in frequency; Sk. to Jmt., also Nb. — Norway: separate distributions in the south and in the extreme north. — Finland: generally distributed, but local; also in the adjacent parts of the USSR. — Europe except in the south.

Biology. A sun-loving species, characteristic of *Calluna*-vegetation. It occurs on dry heaths, e.g. on the "alvar" steppe of Öl. and Gtl., as well as in wet habitats like peat bogs. *C. nitens* is a diurnal spring breeder, most numerous in May, and shows very little autumn activity.

25. *Carabus convexus* Fabricius, 1775
Pl. 2: 3.

Carabus convexus Fabricius, 1775, Syst. Ent.: 238.

16-19 mm. Together with *nitens* the smallest member of the genus, characterized by the extremely fine and regular elytral sculpture. Black, pronotum and outermost margin of elytra usually with a faint steel-blue hue. Elytra proportionally broader than in all other species, with a system of very dense longitudinal ridges and 3 almost inconspicuous rows of punctures. Male with terminal segment of palpi more axe-shaped.

Distribution. Denmark: widely distributed but rare on the islands; in eastern Jutland north to Randers fjord; no records from WJ, NWJ, NEJ and B. — Sweden: in the south-western districts from Sk. to Dlsl., Ög.; rare, especially after 1950. — Norway: in the extreme south-eastern corner. — Finland: in the extreme south, many records from the Hangö peninsula; also Vib. — Europe to West Siberia.

Biology. The species prefers open, warm and dry habitats, such as sun-exposed hillsides. It occurs on clayey as well as on sandy or gravelly, humus-mixed soil. Notably on grassland, but also on heather-covered ground and in cultivated fields. Reproduction takes place in spring, and newly emerged beetles occur in August-September.

26. *Carabus intricatus* Linnaeus, 1761
Pl. 2: 7.

Carabus intricatus Linnaeus, 1761, Fauna Suec. ed. 2: 217.

25-38 mm. Very characteristic because of the strongly prolonged head, the narrow almost quadrangular pronotum, and the rough elytral sculpture. Appendages very long and slender. Black, pronotum and elytra bluish or violaceous, at least laterally. Hind-angles of pronotum acute, protruding. Main carinae as well as elytral intervals dissolved into tubercles. Male with terminal segment of palpi more axe-shaped than in the female.

Distribution. Everywhere local and rare. Denmark: only in EJ (Vejle, Silkeborg and Hald) and B. — In Sweden only in the south-east corner of Sk. (Stenshuvud, Kivik, Forsaker, repeatedly found after 1950). — Otherwise lacking in Fennoscandia. — Europe except in the north and south-west.

Biology. In Scandinavia confined to deciduous ravine forests, usually beech forests, occurring in moist, shaded habitats rich in litter. In C. Europe often in mountain forests. Reproduction takes place in spring. The adults hibernate in decaying tree stumps, fallen trunks, etc.

Note. The species was mentioned from Sweden by Linnaeus (1761), but in recent time it was unknown from this country until 1942, when found in Skåne. It is now a protected species in Sweden.

27. *Carabus violaceus* Linnaeus, 1758
   Fig. 66; pl. 2: 4.

Carabus violaceus Linnaeus, 1758, Syst. Nat. ed. 10: 414.

20-30 mm. Agreeing with *glabratus* in possessing an irregular, almost smooth elytral sculpture (see no. 22 for separating characters). Dull black, almost constantly with a metallic (violaceous or blue, rarely greenish or golden) lustre along the side-margins. Penultimate segment of labial palpi with several setae. Elytra almost smooth with minute granulae, which are either quite irregularly distributed or joined into 3 faint longitudinal ridges. Rufinistic individuals are rare.

Distribution. Denmark: widely distributed and rather common, but seemingly absent from parts of western and northern Jutland, and from several of the islands, e.g. Falster and Møn. — Sweden, Norway and Finland: generally distributed. — Europe to E. Siberia and Japan.

Biology. A eurytopic forest species preferring light deciduous or coniferous forest on sandy as well as clayey, rather dry soil rich in humus. To a lesser extent in shady habitats in open country; in the alpine region in densely vegetated sites. It is a nocturnal species. The newly emerged adults occur in summer and reproduce mainly in August. Some adults hibernate and enter upon a second breeding in the following summer.

Note. This is a multiform species with many subspecies recognized on the continent, in part difficult to separate from *purpurescens* F. The Fennoscandian population has been differently divided by Born (1926) and Mandl (1962). According to the latter author, the Fennoscandian forms should be divided in such originating from glacial refugia on the Norwegian coast (*ottonis* Csiki, 1909; *arcticus* Sparre Schneider, 1888; *pseudoarcticus* Mandl, 1962) and the postglacial immigrants from the continent *(violaceus* L. s. str.; *lindbergi* Burkhart, 1921; *carelicus* Hellén, 1936). Transitional forms between these populations are so frequent that the maintainance of subspecific status seems unnecessary.

28. *Carabus coriaceus* Linnaeus, 1758
Pl. 2: 6.

Carabus coriaceus Linnaeus, 1758; Syst. Nat. ed. 10: 413.

32-40 mm. This, the largest of all our Carabidae, has often been placed in a separate genus *Procrustes* Bonelli, on account of the irregularly reticulate elytral sculpture, the medially produced labrum, the lack of setae on first antennal segment, etc. On the other hand, it seems nearly related to *violaceus,* with which it is able to hybridize.

Coal black. The elytral sculpture consists of a net-work of wrinkles, which may fuse as to form 2 or 3 suggested longitudinal lines. Male with only 3 dilated pro-tarsal segments.

Distribution. Denmark: common in eastern Jutland, and in NEZ, SZ and LFM; only isolated localities in WJ, NWJ, NEJ, F; not found on B. — Sweden: Sk.-Hls., but with large distributional gaps. — Norway: along the coast to Nsy, Vega (J. Andersen). — Not found in Finland, but occurs in the Leningrad-area. — Main part of Europe, but not on the British Isles or the Iberian Peninsula; Asia Minor.

Biology. In southern Sweden almost confined to beech forest which is also the preferred habitat in Denmark. Elsewhere the species is more eurytopic, occurring in both deciduous and coniferous stands, on soil rich in humus and with a moderate content of moisture. The species is strictly nocturnal. Newly emerged beetles occur in late summer and reproduce in the autumn, mainly in August-September. A number of adults hibernate and become active in the following spring. Later they enter upon an aestivation dormancy. Some reproduce for a second time during the autumn. In northern and alpine districts development lasts two or three years.

Tribe Cychrini

The habits as snail-feeders has marked the members of this tribe, above all through their narrow, prolonged head with prominent bidentate mandibles. The elytra are inflated, coalescent along the suture, and the hind-wings are virtually totally reduced.

The tribe is poorly represented in the Old World and only the nominate genus occurs in our area. The North American fauna contains many more forms.

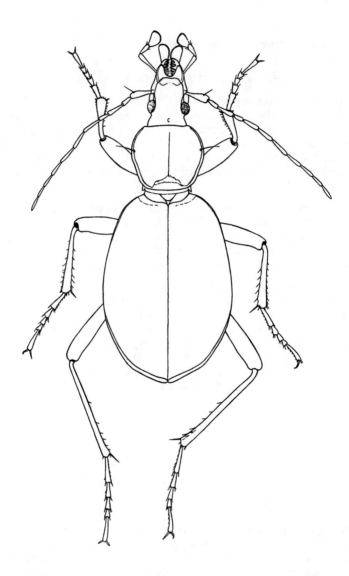

Fig. 71. *Cychrus caraboides* (L.), length 14-19 mm. Surface sculpture omitted.

Genus *Cychrus* Fabricius, 1794

Cychrus Fabricius, 1794, Ent. Syst. 4 App.: 440.
  Type-species: *Tenebrio caraboides* Linnaeus, 1758.

Unmetallic black. Pronotum narrow, oblong. Elytra oviform, very convex. Labrum deeply bilobed.

29. *Cychrus caraboides* (Linnaeus, 1758)
Fig. 71; pl. 3: 1.

Tenebrio caraboides Linnaeus, 1758; Syst. Nat. ed. 10: 418.
Tenebrio rostratus Linnaeus, 1761; Fauna Suec. ed. 2: 283.

14-19 mm. Pronotum rugosely punctate, elytra with dense granulae, which may be arranged so as to indicate 2 or 3 longitudinal ridges. Male with terminal palpal segments axe-shaped.

The species has been divided into 2 (or more) varieties: (a) forma *rostratus*. With a more opaque lustre, better defined hind-angles of pronotum, no (or only slight) tendency of elytral ridges.

(b) forma *caraboides* (incl. *pygmaeus* Chaudoir and *convexus* Heer). Smaller on an average, more shiny due to a weaker microsculpture, and usually with indication of 2-3 elytral ridges.

Forma *rostratus* has a more southern distribution than forma *caraboides*. All transition forms between them occur, wherefore they do not deserve subspecific status.

Distribution. Generally distributed and common in most parts of Denmark and Fennoscandia. — A strictly European species, eastward to E. Russia.

Biology. Predominantly a woodland species, preferring deciduous forest on shady, rather moist soil rich in humus. Often in tree-stumps and fallen trunks; in mountainous regions also in open country above the timber line. The species has a well-known ability to stridulate when disturbed. The sound is produced by rubbing stridulatory surfaces at the edges of abdomen against a flange inside the lateral edges of the elytra. This behaviour is probably a warming signal and is often associated with the ejection of an acid secretion from abdominal glands (Claridge 1974). When feeding on snails (see above), *Cychrus* penetrates into the shell with its head and forebody; also slugs are fed upon. The prey is digested extra-intestinally. It is a strictly nocturnal species. Newly emerged beetles occur in summer and reproduce in the autumn. Not only the larvae, but also a number of adults hibernate and become active in the following spring.

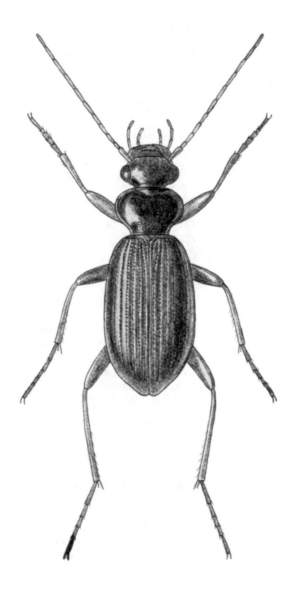

Fig. 72. *Leistus ferrugineus* (L.), length 6.5-8 mm. (After Victor Hansen).

Tribe Nebriini
Genus *Leistus* Fröhlich, 1799

Leistus Fröhlich, 1799, Naturforscher Halle 28: 1.
Type-species: *Carabus ferrugineus* Linnaeus, 1758.
Pogonophorus Latreille, 1802, Hist. Nat. Crust. Ins. 3: 88.
Type-species: *Carabus spinibarbis* Fabricius, 1775: 243.

Rather distantly related to the two following genera and easily distinguished by the very flat dilated madibles, the spiny lateral edges of the maxillae, the slender palpi, and the constricted neck of the head (Fig. 73). Pronotum with or without a seta at hind-angle. Hind-wing varying in size, even within the same species *(rufomarginatus)* and often non-functional. Male with 3 dilated pro-tarsal segments. Penis quite different from that of *Nebria* and much varying between species.

The species of this genus occur among debris in more or less shady places, where they prey upon mites, Collembola, etc. They are nocturnal.

Key to species of *Leistus*

1 Upper surface dark brown with paler margins on pronotum
   and elytra. Pronotum with a seta at hind-angle. Elytral
   shoulders with a small tooth . . . . 30. *rufomarginatus* (Duftschmid)
- Upper surface either entirely pale (rarely brown) or with at
   most head, apex of elytra and suture darker. Pronotum
   without latero-basal seta. Shoulders without tooth 2
2(1) Unicolorous yellow or brown. Pronotum (Fig. 74) with right
   hind-angles and sides in front of these parallel for a short
   distance . . . . 32. *ferrugineus* (Linnaeus)
- Yellow to brown with head and abdomen black, usually also
   apex and suture of elytra dark. Sides of pronotum (Fig. 75)
   diverging from the obtuse hind-angles . . . . . 31. *terminatus* (Hellwig *in* Panzer)

30. *Leistus rufomarginatus* (Duftschmid, 1812)
   Pl. 4: 3.

Carabus rufomarginatus Duftschmid, 1812, Fauna Austriae 2: 54.

8-9.5 mm. The pale margins of pronotum and elytra are characteristic of this species. Broad and flat with elytra parallel-sided at middle. Dark brown to piceous, underside somewhat paler, margins reddish yellow. Separated from the two following species also by the flattened side-margin of pronotum and the elytral striae being sharply incised to apex.

Distribution. Denmark: generally distributed and often abundant; less so in WJ and

NWJ. — Sweden: Sk. (common), Bl., Hall., Vg. — Not in Norway or Finland. — Eastern part of C. Europe. The species has expanded considerably towards the west and north during the last decades, including to the British Isles and southernmost Sweden.

Biology. Mainly in deciduous forest on mull soil with a moderate content of moisture, notably in light stands of beech, but also in darker forest. Adults emerge in late June and then aestivate, often aggregated, under bark of tree stumps or under logs lying on the forest floor. They retain activity and reproduce in late autumn. Only a few adults seem to survive the following winter.

31. *Leistus terminatus* (Hellwig *in* Panzer, 1793)
Figs 73, 75.

Carabus rufescens Fabricius, 1775, Syst. Ent.: 247 (preocc. by *Carabus rufescens* Ström, 1768).
Carabus terminatus Hellwig *in* Panzer, 1793, Fauna Ins. Germ. 7: t. 2.

6-8 mm. A convex species with pronouncedly oviform elytra. Constantly darker than *ferrugineus* (see the key), even the palest specimen has at least head and abdomen black, the darkest specimens are almost uniformly piceous. Separated from *ferrugineus* also by the form of the pronotum (Fig. 75), the flatter eyes and the more convex elytra.

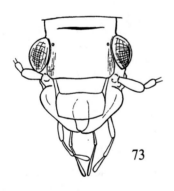
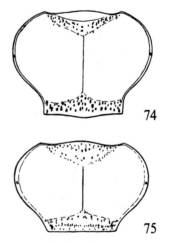

Figs 73-75. *Leistus.* — 73: head of *terminatus* (Hellw. in Pz.); 74: pronotum of *ferrugineus* (L.); 75: pronotum of *terminatus* (Hellw. *in* Pz.).

Distribution. In Denmark widely distributed and common. — Sweden: common and generally distributed except in parts of Lapland. — Norway: to the Polar Circle, also in some places further north. — Not common in southern and central parts of Finland (north to ObS) and adjacent parts of the USSR. — Europe except the SW, S and SE parts, over Siberia to the Pacific Ocean.

Biology. The most hygrophilous member of the genus, living among wet leaves in moist, shady sites in hardwood forest, for instance along brooks in ravines and in alder-swamps. An autumn breeder. The life cycle is apparently similar to that of *L. rufomarginatus,* and the species has the same tendency to form aggregations.

32. *Leistus ferrugineus* (Linnaeus, 1758)
    Figs 72, 74; pl. 4: 4.

Carabus ferrugineus Linnaeus, 1758, Syst. Nat. ed. 10: 415.

6.5-8 mm. Somewhat narrower and less convex than *terminatus.* Unicolorous, usually yellow, rarely brown to almost piceous. The shape of the pronotum is the best distinguishing character (Fig. 74).

Distribution. Denmark: rather well distributed and rather common. — Sweden: generally distributed and common in the southern half of the country, in the north only a few localities in Jmt. and Lpm. — Norway: along the coast to the northernmost parts. — Finland: rather common in the southern parts, north to 64°N; also Vib and Kr in the USSR. — Almost entire Europe except the southern parts.

Biology. More xerophilous than the two preceding species, occurring in open country with rather dense vegetation of grasses etc., in hedges, and in open woodland; always on quite dry, sandy or gravelly soil. Mostly solitary. It is an autumn breeder having an aestivation period as adult.

Genus *Nebria* Latreille, 1802

Nebria Latreille, 1802, Hist. Nat. Crust. Ins. 3: 89.
   Type-species: *Carabus brevicollis* Fabricius, 1792.
Helobia Stephens, 1828, Ill. Brit. Ent. Mand. 1: 60.
   Type-species: *Carabus brevicollis* Fabricius, 1792.

The most characteristic features are the short, cordiform pronotum and the very long and slender appendages (Fig. 76). Our species have two lateral setae on pronotum, the posterior one at hind-angle. The four basal segments of antennae are without pubescence. Elytra with 9 complete striae plus one abbreviated stria at scutellum; 3rd interval (sometimes also 5th and 7th) with a few dorsal punctures. Male with 3 dilated protarsal segments.

A large genus with a circumpolar distribution, especially rich in species in the mountains of C. Europe and western N. America.

They are carnivorous beetles, preying mostly upon small arthropods like Diptera-larvae, Collembola and mites. The ability of flight is not well known.

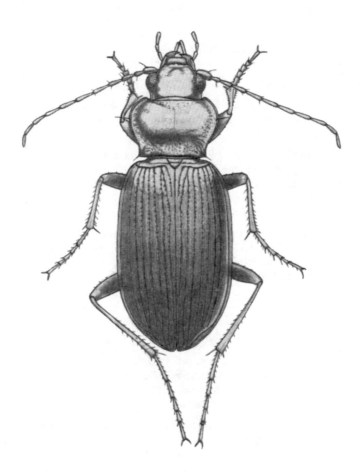

Fig. 76. *Nebria brevicollis* (F.), length 10-14 mm.

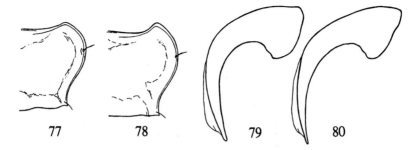

77 78 79 80

Figs 77, 78. Pronotum of *Nebria*. — 77: *nivalis* (Payk.); 78: *rufescens* (Strøm).
Figs 79, 80. Penis of *Nebria*. — 79: *brevicollis* (F.); 80: *salina* (Fairm. & Lab.).

Key to species of *Nebria*

1   Pronotum pale. Elytra bicoloured, black with sides and apex
more or less broadly yellowish . . . . 33. *livida* (Linnaeus)

–   Pronotum and elytra unicolorous, brownish to blackish, or
elytra rufinistic 2

2(1) Shoulder-angle of elytra sharp (Fig. 76). Antennae entirely pale 3

–   Shoulder-angle obtuse or rounded. Antennae black or pice-
ous, sometimes with basal segments slightly paler 4

3(2) Meso- and meta-tarsi with basal segments finely pubescent
above. Elytral microsculpture consisting of transverse meshes,
at least twice as broad as long . . . . 36. *brevicollis* (Fabricius)

–   All tarsi (except for apical setae on each segment) glabrous
above. Meshes of elytral microsculpture only slightly trans-
verse . . . . 37. *salina* Fairmaire & Laboulbène

4(2) No ridge present inside anterior seta of pronotum (Fig. 78).
Femora dark, tibiae brown to piceous, or legs entirely ru-
fous. Only third elytral interval with dorsal punctures . . 34. *rufescens* (Ström)

–   Anterior marginal seta of pronotum inwardly accompanied
by a minute ridge (Fig. 77). Femora rufous with dark apex
(or pale at base only); tibiae nearly or totally black. Elytra
often with dorsal punctures both on 3rd and 5th intervals 35. *nivalis* (Paykull)

33. ***Nebria livida*** (Linnaeus, 1758)
    Pl. 3: 2.

Carabus lividus Linnaeus, 1758, Syst. Nat. ed. 10: 414.

14-16 mm. Our largest species, easily recognized by the contrasting coloration.
Ground colour dark, black to piceous, frons with two red spots, pronotum (except

base and anterior margin), elytra along sides and at apex, and all appendages, testaceous (very pale in live specimens). Male with broader, somewhat arcuate meso-tibiae.

Distribution. Danmark: rare and very local, only rather common in NEZ and B; NWJ (Fur), NEJ (Lønstrup), F (Ærø); not known from SZ and Lolland. — Sweden: not common, north to 60°N, mostly in the west. — Norway: local in the south-eastern parts. — Finland: rare, north to 62°N; also in southern parts of Kr and in Vib. — C. and E.Europe, over Siberia to Japan.

Biology. On wet, sterile banks and shores consisting of sand or sand mixed with clay. In Fennoscandia usually on lake-shores and river banks, occasionally on the seashore. Often together with *Omophron limbatum*. In Denmark it is almost restricted to clayish sea-slopes, but is also found inland in old lignite beds. The species is strictly nocturnal and hides during daytime in clay cracks, under refuse, etc. Newly emerged adults occur in spring, and reproduction takes place in the autumn.

34. *Nebria rufescens* (Strøm, 1768)
Fig. 78.

Carabus rufescens Ström, 1768, Skr. K. Norske Vidensk. Selsk. 4: no. 32.
Carabus Gyllenhali Schönherr, 1806, Syn. Ins. 1: 196.
Nebria hyperborea Gyllenhal, 1827, Ins. Suec. 4: 415.

9-13.5 mm. This and the following species are smaller and more slender than *livida* and *brevicollis*. Black, in northern populations often with more or less rufinistic elytra (*rufescens* s.str.). Palps, base of antennae, tibiae and tarsi normally piceous; a form has legs, and usually also antennae and mouth-parts, clearly rufous. Pronotum: fig 78. Shoulder-angle of elytra obtuse but not quite rounded. Only 3rd interval with a few, non-foveate dorsal punctures. Last abdominal sternite with microsculptural meshes arranged in transverse rows.

Distribution. Not in Denmark. — Sweden: generally distributed and common in the north, south to the large lakes in the central part, and to Öl. and Gtl. — Norway: entire country. — Finland: widely distributed but local, west to Al; common in the north. Also generally distributed in the Soviet part of E. Fennoscandia. — Circumpolar, represented by different subspecies in E. Asia and N. America. In Europe boreoalpine.

Biology. A cold-loving species, preferring temperatures at about 8°C (Krogerus, 1960). It is especially prominent in the lower alpine region where it is rather ubiquitous, often occurring along glacial streams. At lower altitudes it is more or less confined to stony or gravelly shores of rivers, brooks and lakes with cold water; in Norway quite eurytopic also in the lowland (Andersen, 1982). In southern Fennoscandia the adults emerge in early summer and reproduce in the autumn. In northern and alpine regions the life cycle is biennial.

35. *Nebria nivalis* (Paykull, 1790)
   Fig. 77.

Carabus nivalis Paykull, 1790, Mon. Car. Suec.: 184.

9-11 mm. Similar to the normal dark form of *rufescens* but separated in the following points: body narrower, somewhat more convex. Femora normally rufous (rarely pale at base only), tibiae black or almost black. Pronotum: fig. 77. Elytra with more rounded shouldrs, striae usually finer and more evidently punctate. In addition to the dorsal puncture on 3rd interval usually with one or more punctures also on 5th (sometimes also on 7th) intervals, all more or less foveate (in *rufescens* at most suggested).

Distribution. This is the most pronouncedly arctic element among all Fennoscandian Carabidae, and records from localities below the timber limit are probably accidental. — Not in Denmark. — Sweden: Jmt.-Lpl., many localities north of 66°N. — Norway: in the southern mountains and north of the Polar Circle. — Finland: Lapland, locally abundant. — Also Lr in the USSR. — Circumpolar in high latitudes: W. Siberia, N. America, Scotland.

Biology. This species prefers even lower temperatures than *N. rufescens* (Krogerus, 1960) and is most numerous in the upper alpine zone. The habitat is poor high mountain grounds, usually near the border of glaciers, along glacial streams, and on the shores of lakes with cold water, often on stoney or gravelly banks. It is sometimes found on snow fields in search for frozen insects. Mainly in June-August.

36. *Nebria brevicollis* (Fabricius, 1792)
   Figs 76, 79; pl. 4: 5.

Carabus brevicollis Fabricius, 1792, Ent. Syst. 1: 150.

10-14 mm. Piceous or dark brown, extreme sides of pronotum and elytra somewhat translucent, appendages dark rufous but femora (in exceptional cases also base of antennae) darker. Raised marginal bead of pronotum thick. Elytral striae coarse, strongly punctate, intervals more or less convex. Elytral microsculpture consisting of transverse meshes, at least twice as broad as long. Meso- and meta-tibiae with basal segments finely pubescent above. Penis (Fig. 79) evenly arcuate.

Distribution. Denmark: common everywhere. — Sweden: common and widespread from Sk. to S. Upl. — Norway: along the coast to 62°N. — Finland: only Al, many localities. — Europe and W. Asia.

Biology. A eurytopic woodland species, predominantly in deciduous forests, in Scandinavia mainly of beech, on moist mull soil; also on shady ground in open country, for instance in parks and gardens. It is nocturnal and an autumn breeder. Newly emerged adults occur in spring. After a short period of activity they enter upon summer dormancy, during which the beetles aggregate under bark of tree stumps, under logs in the forest floor, etc. Activity is resumed in the autumn when breeding takes place. Larvae and a number of adults hibernate.

37. *Nebria salina* Fairmaire & Laboulbène, 1854
   Fig. 80.

Nebria salina Fairmaire & Laboulbène, 1854, Faune Ent. France 1: 14.
Nebria degenerata Schaufuss, 1862, Rev. Zool. 1862: 491.
Nebria Klinkowstroemi Mjöberg, 1915, Ent. Tidskr. 36: 285.

10-13.5 mm. This species was long confused with *brevicollis* but differs in the following respect: the body is slightly flatter and narrower, the elytra more parallel-sided, and upper surface more shiny. Same coloration, except that penultimate segment of maxillary palpi is infuscated. Pronotum with a narrower lateral bead and base slightly more constricted. Elytral striae usually finer and more faintly punctate. The most reliable character is the absence of a tarsal pubescence. Penis (Fig. 80) more slender and less arcuate.

Distribution. Denmark: widely distributed but rather local. — Sweden: distributed from Sk. to Boh. & Gtl.; local, common only on Öl. and Gtl. — Norway: along the coast to 64°N. — Not in East Fennoscandia. — N. and W. Europe. Originally strictly Atlantic, but has spread to the inland of C. Europe in the last few decades.

Biology. Usually inhabiting drier and more open country than the preceding species. It prefers clay-mixed gravel and sand, but also occurs on peaty soil, for instance in gravel pits, clay pits and moist places in dunes; also on arable land. Occasionally together with *N. brevicollis*, e.g. at forest edges. It is nocturnal. The life cycle is similar to that of *N. brevicollis*.

Genus *Pelophila* Dejean, 1821

Pelophila Dejean, 1821, Cat. Coll. Col. B. Dejean: 7.
Type-species: *Carabus borealis* Paykull, 1790.

General habitus as in *Nebria* but appendages shorter. Easily separated, also from *Blethisa* with a similar elytral sculpture, by the presence of 10 complete striae (no. 2, corresponding to the "scutellar" one, only slightly abbreviated apically); 4th and 6th intervals with an irregular row of foveae. Male with 3 strongly dilated pro-tarsal segments.
Only one species.

38. *Pelophila borealis* (Paykull, 1790)
   Pl. 4: 6.

Carabus borealis Paykull, 1790, Mon. Car. Suec.: 61.

9-12.5 mm. Piceous to black, upper surface almost constantly with a metallic, brassy, rarely greenish or bluish, reflection; seemingly black individuals retain the metallic hue at least on the edges of pronotum and elytra; a variety with rufinistic elytra is rare in our area. Appendages black to brown, sometimes legs rufous with dark knees.

Distribution. Not in Denmark. — In Fennoscandia in the conifer region south to 60-62°N and far north of the Polar Circle; fairly common in the north. — Circumpolar; in Europe outside Fennoscandia and N. Russia only on the British Isles.

Biology. This species occurs from the conifer region up to the lower alpine region, living at the margins of lakes and slow-running rivers, in places where the soil is silty or muddy and vegetated by *Carex, Eriophorum,* etc. It is especially abundant in the lower and middle section of the large rivers in the north, often living in association with *Agonum dolens*. The adults hibernate in cavities in the soil which are sometimes filled with water during autumn flooding (Östbye & Sömme, 1972), and the beetles are subsequently often enclosed by ice. The survival of *P. borealis* under these conditions is shown by Conradi-Larsen & Sömme (1973) to be dependent on anaerobic metabolism, by which lactate accumulates in the haemolymph, and on the ability to withstand freezing at least at temperatures down to -5°C (Sömme, 1974). The species occurs from late May to early September; newly emerged imagines have been found in July-August.

Tribe Notiophilini
Genus *Notiophilus* Duméril, 1806

Notiophilus Duméril, 1806, Zool. An., Paris: 194.
Type-species: *Cicindela aquatica* Linnaeus, 1758.

One of the most characteristic carabid genera (Fig. 81), easily recognized by the enormous eyes and the furrowed frons (Figs 82, 83), as well as by the broad second elytral interval.

They are small beetles (less than 7 mm), with narrow, parallel-sided elytra. These with strong, punctate striae; 4th interval with at least one dorsal puncture and one or two preapical punctures (Figs 85, 86). Many species exhibit wing dimorphism, the brachypterous form having only a scale-like remain. Male with terminal segment of both palpi more or less dilated; also 3 pro-tarsal and one meso-tarsal segments slightly widened.

The species are diurnal, sun-loving insects, very rapid in their movements. They are visually hunting beetles, preying upon mites, Collembola and other arthropods. The larvae have specialized on Collembola to a greater extent than have the adults.

Key to species of *Notiophilus*

1 2nd elytral interval just behind middle more than 3 times
   as wide as 3rd interval 2
– 2nd elytral interval about twice as wide as 3rd interval 4
2(1) Elytra uniformly dark. Legs bright rufous, femora of-

ten somewhat infuscated .... 43. *rufipes* Curtis
- Elytra with large yellow apical spot. Legs blackish, only tibiae more or less pale 3
3(2) Apical elytral spot extending almost to shoulder. 7th stria just behind middle weaker than 6th .... 44. *reitteri* Spaeth
- Apical elytral spot not prolonged, or only to just before middle. 6th and 7th striae equally strong .... 45. *biguttatus* (Fabricius)
4(1) Legs entirely black. Pronotum narrower (Fig. 84) 5

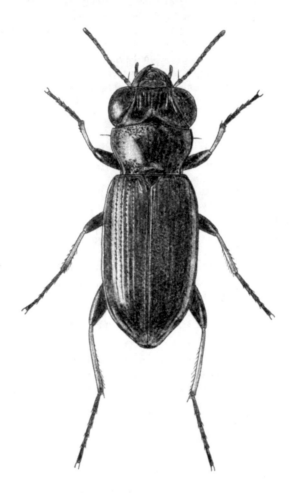

Fig. 81. *Notiophilus palustris* (Dft.), length 5-6 mm. (After Victor Hansen).

–   Tibiae more or less pale, at least at apex. Pronotum broader
     with more rounded sides 6
5(4) Elytra with one preapical puncture (Fig. 85) (a rudimentary
     anterior puncture exceptionally present). Elytral intervals
     3-7 quite smooth . . . . 40. *aquaticus* (Linnaeus)
–   Elytra with two preapical punctures (Fig. 86). Elytral inter-
     vals 3-7 with irregular row of very small, flat impressi-
     ons . . . . 39. *aesthuans* Motschulsky
6(4) Frontal furrows parallel or almost so (Fig. 83). Outer elytral
     intervals dull from dense micro-reticulation . . . . 42. *germinyi* Fauvel
–   Frontal furrows forwardly diverging (Fig. 82). Elytral inter-
     vals smooth, shiny . . . . 41. *palustris* (Duftschmid)

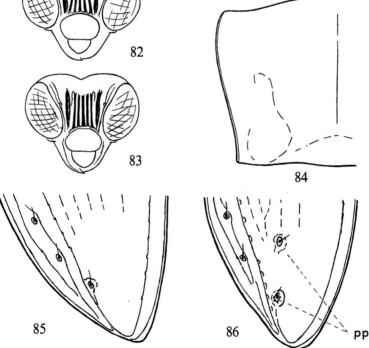

Figs 82, 83. Heads of *Notiophilus*. — 82: *palustris* (Dft.); 83: *germinyi* Fauv.
Fig. 84. Pronotum of *N. aquaticus* (L.).
Figs 85, 86. Apex of elytron of *Notiophilus*. — 85: species with one, and 86: species with two
preapical punctures (= pp).

39. *Notiophilus aestuans* Motschulsky, 1864

Notiophilus aestuans Motschulsky, 1864, Bull. Soc. Nat. Mosc. 37: 164.
Notiophilus pusillus Waterhouse, 1833, Ent. Mag. 1: 207; *nec* (Schreber, 1759).

4-5.5 mm. Difficult to separate from small specimens of *aquaticus*. Somewhat narrower and with flatter elytra (viewed in profile). Punctuation of pronotum finer and denser. The elytral striae reach nearer to apex; the micro-impressions on outer intervals (see the key) are quite shallow and irregular; the presence of two preapical punctures is a constant feature (as in fig. 86). As far as known this species is constantly macropterous. Male with terminal segment of labial palpi dilated, axe-shaped.

Distribution. Denmark: widely distributed but very local and rare; apparently absent from W. and NW. Jutland. — Sweden: very local and rather rare from Sk. to Hls. — Norway: along the coast in the extreme southern part. — Finland: rather rare, north to 62°N; also Vib and the Svir area in the USSR. — Main part of Europe, W. Asia.

Biology. A rather xerophilous species, living in open, dry country on gravelly and sandy soil, often mixed with clay; for instance on *Calluna* heaths, on sandy grassland (e.g. Corynephoretum), in gravel pits, and on sandy lake shores in good distance of the water. It is most numerous in spring and summer.

40. *Notiophilus aquaticus* (Linnaeus, 1758)
   Fig. 84.

Cicindela aquatica Linnaeus, 1758, Syst. Nat. ed. 10: 408.

4.5-6 mm. Narrow but with rather convex elytra. Black, lower surface with a faint, upper surface with a stronger metallic lustre; this may be brass, greenish, or rarely bluish. Base of antennae (at least underneath) and palpi more or less pale. Single specimens may have the elytra slightly and diffusely rufinistic, notably apically. Head as wide as pronotum, frontal furrows parallel. Pronotum (Fig. 84) with sides slightly rounded. Elytral striae moderately incised. One preapical puncture (as in fig. 85). The species is dimorphic, but short-winged specimens are dominating in most populations. Male with almost imperceptably dilated palpi.

Distribution. Generally distributed in all four countries, though seldom abundant. This is the widest distributed carabid beetle in Fennoscandia. — Circumpolar, though clearly multiformous; a division into subspecies may be possible in the future.

Biology. In different kinds of open, dry country, mostly on gravel. In southern Scandinavia occurring on grassland, in forest edges, and on *Calluna* heaths with scattered pine growth. In the mountainous regions mainly on heaths with grasses or dwarf shrubs, e.g. *Empetrum,* often in association with *Amara alpina.* From early spring to late autumn.

41. *Notiophilus palustris* (Duftschmid, 1812)
Figs 81, 82.

Elaphrus palustris Duftschmid, 1812, Fauna Austriae 2: 192.

5-6 mm. Easily recognized by the anterior diverging frontal furrows (Fig. 82). Broad and shiny. Upper surface brassy (exceptionally bluish), frons often with a more varying lustre in red and green. Tibiae, base of palpi and 4 basal segments of antennae pale. Eyes larger than in any other species, head therefore wider than pronotum, which has strongly rounded sides. Elytral intervals not reticulate, striae dense and deep, especially behind shoulder. Dimorphic, but long-winged specimens are rare.

Distribution. Denmark: generally distributed and common. — Sweden: generally distributed and common in the south, north to the Polar Circle. - Norway: along the southern coast to SFy. — Finland: northwards to the Polar Circle; also Vib and Kr. — Almost entire Europe, except northernmost parts; W. Asia.

Biology. A hygrophilous species, occurring in rather shady localities on humus-rich soil. Found in deciduous woodland among litter and mosses, e.g. in the drier places of *Alnus glutinosa* swamps, as well as in open country such as meadows and marshes with high and rather dense vegetation. Most numerous in spring when breeding takes place, but also in autumn when the new generation of adults emerges.

42. *Notiophilus germinyi* Fauvel, 1863
Fig. 83.

Notiophilus germinyi Fauvel, 1863, *in* Grenier: Cat. Col. France 1: 1.
Notiophilus hypocrita auctt.

5.5-6.5 mm. Related to *palustris* but differing in the following points: narrower, notably the head, which hardly exceeds the width of pronotum; this has the sides less rounded anteriorly. Frontal furrows virtually parallel (Fig. 83). Metallic lustre more vivid, often greenish along elytral sides. Striae weaker. Constantly macropterous.

Distribution. Denmark: generally distributed and rather common. — Sweden and Norway: all over the countries. — East Fennoscandia: generally distributed but more local. — Almost entire Europe.

Biology. A xerophilous species inhabiting different kinds of more or less open, dry country such as sandy grass areas; often together with *N. aquaticus*. In Denmark especially found in *Calluna* vegetation both on open heather (which is the favoured habitat in N. Germany) and in thin coniferous forest. In the Scandinavian mountains mainly in dry birch forest, but also in the lower parts of the alpine zone. It is an autumn breeder; most numerous in June-August.

43. *Notiophilus rufipes* Curtis, 1829

Notiophilus rufipes Curtis, 1829, Brit. Ent. 6: 254.

5.5-6.5 mm. Similar to *palustris* in general outline but with more stretched elytra. Upper surface with a strong brassy lustre. Legs bright rufous, often with femora and apex of each tarsal segment infuscated; base of antennae and palpi paler than pronotum, frontal furrows slightly diverging anteriorly. Elytral striae more coarsely punctate than in *palustris,* 2nd interval broad as in *biguttatus.* As far as known constantly macropterous.

Distribution. Denmark: in eastern part of Jutland north to Randers; also WJ (Esbjerg); several localities on the islands; absent from B. — Sweden: rare, only a few records from southern Sk. — Lacking in the rest of Fennoscandia. — Western half of the European continent; W. Asia.

Biology. A woodland species, living among litter and moss in deciduous forest on somewhat moist and shady ground. A spring breeder; most numerous in the reproductive period and in the autumn when the new beetles emerge.

44. *Notiophilus reitteri* Spaeth, 1899

Notiophilus reitteri Spaeth, 1899, Verh. zool.-bot. Ges. Wien 49: 513, 522.
Notiophilus fasciatus Reitter, 1897, Ent. Nachr. 23: 363; *nec* Mäklin, 1855.

5.5-6.5 mm. As in *biguttatus* with bicoloured elytra, the pale apical spot being prolonged along the margin almost to the shoulder (best seen in frontal view under indirect light). More stretched than *biguttatus.* The lustre of upper surface often silvery. Pronotum less widening anteriorly. Outer interval of elytra somewhat broader and more evidently micro-reticulate; 2nd stria at most faintly undulate before apex; 7th stria shallover than 6th about middle.

Distribution. Not in Denmark. — In Sweden distributed in the north, from Lpl. south to Dlr. and Jmt.; rather rare. — Norway: scattered from the extreme north south to about 60°N. — Finland and adjacent parts of the USSR: very local from north to south. — Not in C. and W. Europe, only N. Europe and N. Asia.

Biology. In the southern parts of Fennoscandia mainly in moist, shady areas in spruce forest, where found under moss and among needles. Further north *N. reitteri* is less stenotopic, in N. Norway favouring light birch-wood. In Finland particularly characteristic of peaty woods (Krogerus, 1960). The species is most numerous in summer and is presumably a spring breeder.

45. *Notiophilus biguttatus* (Fabricius, 1779)
Pl. 4: 7.

Elaphrus biguttatus Fabricius, 1779, Reise Norwegen: 222.

5-6 mm. Elytra with a well defined contrasting yellow spot and thereby at once separated from all preceding species. Metallic lustre brass, seldom bluish, base of palpi and antennae, as well as tibiae, rufous. The elytral spot is in exceptional cases prolonged to just before middle. Somewhat broader and flatter than *aquaticus,* with more parallel-sided elytra. Head hardly broader than pronotum, the sides of which are almost straight. Outer elytral intervals weakly micro-reticulate; 2nd stria with S-shaped bow before apex. Two preapical punctures. Wings dimorphic, populations often mixed.

Distribution. Denmark: very common and well distributed. — Sweden, Norway and Finland: generally distributed except in northernmost parts; also Vib and Kr in the USSR. — Virtually distributed in entire Europe, east to the Caucasus. Introduced in North America.

Biology. A eurytopic woodland species, inhabiting clearings in both deciduous and coniferous forests. It lives among litter on more or less dry, sun-exposed ground with sparse vegetation. In humid, oceanic climate in Norway and Iceland it may occur above the timber limit. Breeding takes place in May-June, and newly emerged beetles occur in summer and autumn.

Tribe Elaphrini
Genus *Blethisa* Bonelli, 1810

Blethisa Bonelli, 1810, Obs. Ent. 1 (Tab. Syn.).
Type-species: *Carabus multipunctatus* Linnaeus, 1758.

Characterized by the very peculiar frontal furrows (Fig. 87) and the elytral sculpture, which is similar to that of *Pelophila.* The sides of the pronotum are much more broadly reflexed than in *Diacheila* and *Elaphrus.* Upper surface with a metallic lustre. The elytral striae are represented by serial punctures, which are somewhat irregular because of the two rows of large foveae on 3rd and 5th intervals and an impression behind the shoulder; base with a raised margin. Wings full. Male with 4 dilated pro-tarsal segments. The penis is enlarged, with a stout, arcuate central stylet. The parameres are fringed with long hairs.

The genus is circumpolar, with 6 species in North America, 4 in Siberia, and 2 in Europe (one in our area).

46. ***Blethisa multipunctata*** (Linnaeus, 1758)
    Fig. 87; pl. 4: 8.

Carabus multipunctatus Linnaeus, 1758, Syst. Nat. ed. 10: 416.

10-13.5 mm. Black with a bronzy lustre, margins of pronotum and elytra usually greenish. Penis with a small, sharp tooth ventrally near apex.

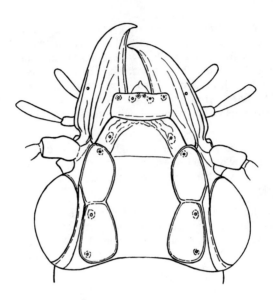

Fig. 87. Head of *Blethisa multipunctata* (L.).

Distribution. Denmark: rare and with scattered distribution. — Sweden: generally distributed, Sk. - T.Lpm. — Norway: entire country except several districts on the west coast at 60-63°N. — Finland: local but generally distributed; also Vib and Kr. — Circumpolar, N. and C. Europe, Siberia, North America.

Biology. A very hygrophilous species, living on marshy soil in fens at the margins of lakes and slowly running rivers, etc., usually in sun-exposed sites with mossy ground. Especially typical of mires with rich vegetation of *Carex* and *Eriophorum,* but not where *Sphagnum* spp. predominate; often in association with *Agonum versutum.* It is a spring breeder, most numerous in May-June.

Genus *Diacheila* Motschulsky, 1844

Diacheila Motschulsky, 1844, Ins. Sibér.: 74.
  Type-species: *Blethisa arctica* Gyllenhal, 1810.
Diachila auctt.
Arctobia Thomson, 1859, Skand. Col. 1: 4.
  Type-species: *Blethisa arctica* Gyllenhal, 1810.

Takes in some respects an intermediate position between *Blethisa* and *Elaphrus.* Terminal palpal segment very long. Head with normal frontal foveae, its punctuation as on pronotum, equally dispersed. The raised margin of pronotum is thin, anteriorly ob-

82

solescent. Elytra with an incomplete basal border, striae represented by more or less irregular rows of punctures, and on third interval several foveate dorsal punctures. Upper surface with a metallic lustre. Male with 4 dilated pro-tarsal segments. Penis of moderate size and central stylet weak.

The three known species are strongly cold-preferent, two of them almost circumpolar, the third restricted to Siberia.

Key to species of *Diacheila*

1 Pronotum (Fig. 88) with a latero-basal carina, sides slightly sinuate in basal half .... 47. *arctica* (Gyllenhal)
- Pronotum (Fig. 89) without lateral carina, sides evidently sinuate before base .... 48. *polita* (Faldermann)

47. *Diacheila arctica* (Gyllenhal, 1810)
Figs 88, 90; pl. 4: 9.

Blethisa arctica Gyllenhal, 1810, Ins. Suec. 2: 96.

7-9 mm. Black with a brassy, coppery or greenish lustre; mandibles and often also tarsi reddish brown. Pronotum (Fig. 88) with a latero-basal carina, sides only slightly sinu-

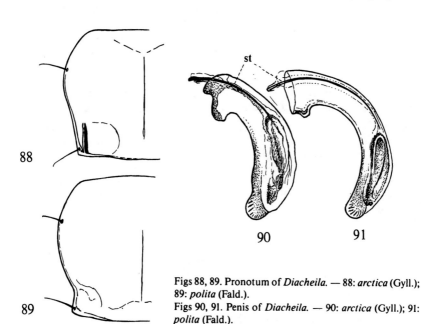

Figs 88, 89. Pronotum of *Diacheila.* — 88: *arctica* (Gyll.); 89: *polita* (Fald.).
Figs 90, 91. Penis of *Diacheila.* — 90: *arctica* (Gyll.); 91: *polita* (Fald.).

ate in basal half. Elytra with regular rows of fine punctures, 5th stria foveate at base inside the protruding shoulders. Elytral microsculpture isodiametric over entire surface. Wings full and functionary. Penis (Fig. 90) stout with a short stylet.

Distribution. Sweden: rare and only found in a few localities in T.Lpm., Lu.Lpm. and P.Lpm. — Norway: Finnmark and one record in Nordland. — In Finland in a few places north of the Polar circle; also Kola Peninsula of the USSR. — Circumpolar, but very scattered.

Biology. A hygrophilous species, living on lake shores and in mires in northernmost Scandinavia, from the upper conifer region to the lower alpine region. The preferred habitat seems to be rich fens with *Carex, Eriophorum, Scirpus, Juncus,* and mosses. At Abisko *D. arctica* occurs in association with *Elaphrus lapponicus* (Palm, 1981). Mostly in spring and early summer; a fully grown larva has been found about mid-July.

48. *Diacheila polita* (Faldermann, 1835)
Figs 89, 91.

Blethisa polita Faldermann, 1835, Mem. Acad. Sci. Math.-Phys. Nat. St. Petersb. 2: 359.

7-8.7 mm. Shorter and more convex than *arctica,* easily recognized by the structure of the pronotum. Black or dark piceous, upper surface more or less brassy. Mandibles, at least bases of antennal segments 3-5, and legs (but usually not femora) paler, reddish. Head narrower than in *arctica,* with less protruding eyes. Pronotum (Fig. 89) without lateral carina, sides evidently sinuate before base. Elytra more widened apically, shoulders not protruding, punctuation coarser and more irregular; also microsculpture more irregular and lacking on disc in the male. Hind-wings reduced into tiny scales. Penis (Fig. 91) slenderer, arcuate, with longer stylet.

Distribution. In Fennoscandia only in the eastern part of the Kola Peninsula (USSR). — On the tundra of northwestern North America and Eurasia. During the Pleistocene glaciation widely distributed in N. & W. Europe.

Biology. Less hygrophilous than *arctica,* usually inhabiting peaty soil on the open tundra. Also at the margin of pools with *Carex,* and sometimes on drier places with *Betula nana.*

Genus *Elaphrus* Fabricius, 1775

Elaphrus Fabricius, 1775, Syst. Ent.: 227.
Type-species: *Cicindela riparia* Linnaeus, 1758.

The elytral sculpture at once separates this genus (Fig. 92) from our other Carabids.

The striae are replaced by alternating rows of shiny rectangular "mirrors" and of pupillate setiferous punctures, each usually surrounded by a depression. Body more or less metallic. Head with enormous, protruding eyes and at least as broad as pronotum, which is coarsely punctate. The raised lateral margin of pronotum is narrow to obsolete. Wings full, only varying in *angusticollis*. Pro-tarsi of male with 3 or 4 dilated segments. The penis is not sclerotized dorsally and contains a heavy stylet as in *Blethisa*.

All *Elaphrus*-spp. are hygrophilous and always occur near water; they are diurnal, visually hunting beetles. When disturbed, *Elaphrus* stridulates by rubbing two rows of bristles on the dorsal side of abdomen against two ridges on the inner side of the elytra, and may at the same time discharge the pygidial defence glands. The species are most numerous in spring and early summer. The biology of *E. cupreus* and *riparius* is treated in detail by Bauer (1974).

Key to species of *Elaphrus*

1   Pupillate punctures of elytra not situated in depressions; punctuation near elytral base tend to be arranged in longitudinal rows. Prosternum with long hairs . . . . 49. *lapponicus* Gyllenhal
-    Pupillate punctures situated in depressions; punctuation of background do not show any longitudinal arrangement. Prosternum glabrous (except in *riparius*) 2

2(1) Elytra shiny, microsculpture weak. Male with 4 dilated pro-tarsal segments 3
-    Elytra (except for "mirrors") quite full from dense microsculpture. Male with 3 dilated pro-tarsal segments 4

3(2) Sides of pronotum strongly rounded, raised lateral bead complete (Fig. 93). All appendages dark metallic . . . . . 50. *uliginosus* Fabricius
-    Sides of pronotum less rounded (Fig. 94), lateral bead obliterating anteriorly. Base of palpi and at least middle of tibiae pale. . . . . 51. *cupreus* Duftschmid

4(2) Pronotum with complete raised lateral bead. Prosternum with long hairs . . . . 52. *riparius* (Linnaeus)
-    Side-margin of pronotum obsolete. Prosternum glabrous. . . . .
. . . . 53. *angusticollis* F. Sahlberg

49. *Elaphrus lapponicus* Gyllenhal, 1810
Pl. 4: 11.

Elaphrus lapponicus Gyllenhal, 1810, Ins. Suec. 2: 8.

8.5-10 mm. A large species, more stretched than all the following, and with narrow shoulders. Quite dull from dense, strong microsculpture. The punctures of elytra in part serially arranged at base, elytral sculpture less developed; the pupillate foveae lit-

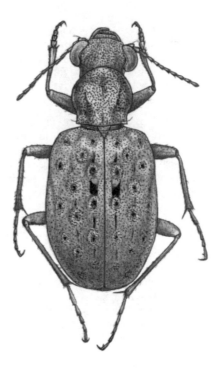

Fig. 92. *Elaphrus riparius* (L.), length 6.5-8 mm.

tle impressed, "mirrors" indistinct. Black with a much varying metallic lustre: coppery, golden, green or bluish; rarely almost black. All appendages dark. Male with 4 dilated pro-tarsal segments.

Distribution. Not in Denmark. — In Sweden distributed but rare and local in the northern half, south to northern Dlr. — Norway: in the mountains north of 60°N; many localities between 66°N and 70°N. — Finland: north of 66°N; also western Kola Peninsula in the USSR. — A circumpolar species in the High North; isolated in N. Britain.

Biology. Predominantly in the birch and the upper conifer region, but also in the lower alpine zone. The species usually inhabits small rich fens with *Carex, Eriophorum,* mosses (e.g. *Paludella*), etc., often near wells and streams. It seems to avoid acid soil. At Abisko it lives in association with *Diacheila arctica.* Although a cold-preferent species, *E. lapponicus* favours higher temperatures (about 13°C) than most other alpine and subalpine species, which agrees with its preference for sun-exposed microhabitats (Krogerus, 1960). It is most numerous in spring and early summer; later in summer only singly.

50. *Elaphrus uliginosus* Fabricius, 1792
   Fig. 93.

Elaphrus uliginosus Fabricius, 1792, Ent. Syst. 1: 178.

8.5-10 mm. Recognized by the proportions of the body: head (incl. eyes) not wider than pronotum, the sides of which are strongly rounded (Fig. 93) and has a complete lateral bead. Upper surface with a variable metallic lustre, usually greenish, but pupillate punctures always more or less violaceous. All appendages dark. Elytra shining, microsculpture weak, punctuation dense, notably around the pupillate punctures, which are situated in evident depressions.

Distribution. Denmark: very local and rare; WJ (Esbjerg, Skallingen), EJ (8 localities at the coast), NEJ (Buderupholm); several records from the islands. After 1950 only 5 records from the whole country. — In Fennoscandia uncommon and very local, northern limit at about 67°N. — Almost entire Europe except the extreme south and north; Asia to E. Siberia.

Biology. On open, very moist soil with mosses and moderately dense vegetation at the margins of lakes, ponds and marshes; often in coastal habitats. It prefers rich fens with *Carex, Eriophorum* and mosses (Krogerus 1960; Jarmer 1973), but also occurs in oligotrophic bogs on acid soil with ericacean shrub, often in association with *Blethisa* and *Agonum versutum*. A spring breeder; predominantly in May-June.

51. *Elaphrus cupreus* Duftschmid, 1812
   Fig. 94.

Elaphrus cupreus Duftschmid, 1812, Fauna Austriae 2: 194.

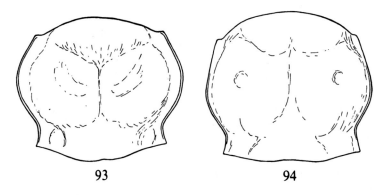

93   **94**

Figs 93, 94. Pronotum of *Elaphrus*. — 93: *uliginosus* (F.); 94: *cupreus* Dft.

8-9.5 mm. Somewhat flatter than *uliginosus* and with more parallel-sided elytra. Pronotum different (Fig. 94), its side-margin more or less obsolete. Dark bronze, sometimes almost black or with a greenish hue; pupillate punctures usually violaceous. Palpi and legs in part pale (see the key), tarsi blue. Elytral microsculpture and punctuation as in *uliginosus*.

Distribution. Generally distributed and common over the entire area, but more scattered in the north-east. — Almost entire Europe; Siberia east to river Lena.

Biology. At the margins of standing or slow-flowing waters, usually on muddy ground with dense vegetation. It is especially typical of eutrophic fens in open country as well as in deciduous woodland; less often found in oligotrophic or even dystrophic bogs. In the mountains it reaches the birch region, but is rare in the alpine zone (Norway, N. Finland). It is a spring breeder, mostly found in May and June.

52. *Elaphrus riparius* (Linnaeus, 1758)
   Fig. 92; pl. 4: 10.

Cicindela riparia Linnaeus, 1758, Syst. Nat. ed. 10: 407.
Elaphrus tuberculatus Mäklin, 1877, Öfvers. Finska Vet. Soc. Förh. 19: 16.
Elaphrus latipennis J. Sahlberg, 1880, K. Svenska Vetenskakad. Handl. 17: 10.
Elaphrus tumidiceps Munster, 1924, Norsk Ent. Tidsskr. 1: 288.

6.5-8 mm. This species is recognized by its small size, and flat and broad elytra which are dulled by dense and strong microsculpture, markedly contrasting against the shiny "mirrors". The hairy prosternum and the presence in the male of only 3 dilated protarsal segments are additional characters. Upper surface usually green, rarely with a yellowish, bluish or bronze lustre, or almost unmetallic. Normally the tibiae (at least the middle), base of femora, and palpi are reddish yellow. Sometimes, especially in northern populations, all appendages are darker, exceptionally entirely metallic; tarsi always green.

Distribution. Denmark and Sweden: common and widely distributed. — Norway: distributed but lacking in some western coastal areas. — Finland and rest of E. Fennoscandia: generally distributed and common, but lacking in the extreme north. — Circumpolar, in Europe south to N. Spain, N. Italy and Serbia.

Biology. Confined to the banks of standing or slow-running waters in open country, mostly on sandy or clayish soil. In the contrary to the preceding species, *E. riparius* prefers sparsely covered or bare sun-exposed ground. It is very active and occasionally flies about in warm sunshine. The species is most numerous in the breeding period in spring.

Note. Henri Goulet (Ottawa) regards *tuberculatus* Mäkl. (with *latipennis* J. Sahlb. and *tumidiceps* Munst. as synonyms) as a separate species. After a careful comparison (Lindroth, 1939) I was unable to draw any clear limit between the northern *"tubercula-*

tus" and the southern *riparius.* That from the north is narrower, generally darker, and elytra less dull because of sparser punctuation, but all transitional stages to *riparius* seem to occur. The inner structure of the penis is also identical. A subspecific separation might be possible. It seems as *"tuberculatus"* has arrived in Fennoscandia via a northern route while typical *riparius* has southern origin.

53. *Elaphrus angusticollis* F. Sahlberg, 1844

Elaphrus angusticollis F. Sahlberg, 1844, Nov. Ochotsk Car. spec.: 20.
Elaphrus angustus Chaudoir, 1850, Bull. Soc. Nat. Mosc. 23 (2): 161.
Elaphrus longicollis J. Sahlberg, 1880, K. Svenska Vetenskakad. Handl. 17: 11.
Elaphrus Jakowlewi Semenov, 1895, Horae Soc. Ent. Ross. 29: 303.

6.3-7.5 mm. More convex than *riparius,* with narrower shoulders. Easily separated on the glabrous prosternum and the obsolete raised margin of pronotum. Upper surface mainly brassy with a greenish or silvery hue. Femora pale, at least basally. Pronotum narrow compared to head. The punctuation of elytra sparser than in normal *riparius.* In contrary to all preceding species *angusticollis* has the hind-wings more or less reduced and apparently not functionary. Male with 3 dilated pro-tarsal segments. Penis strongly and evenly arcuate.

Distribution. In our area only in a few localities in the southernmost of the Soviet part of E. Fennoscandia. — European Russia and throughout Siberia; North America.

Biology. Less hygrophilous than the preceding species, occurring along rivers at some distance from the water. On sand-mixed clay in moderately dense vegetation, usually in rather shaded sites, for instance under bushes. Mostly under subarctic conditions.

Tribe Loricerini
Genus *Loricera* Latreille, 1802

Loricera Latreille, 1802, Hist. Nat. Crust. Ins. 3: 88.
    Type-species: *Carabus pilicornis* Fabricius, 1775.
Lorocera auctt.

This genus has an isolated position. It is characterised by the supernumerous elytral striae and the structure of the antennae. Head (Fig. 95) with very convex eyes, and mandibles explanate laterally (as in *Leistus*). Antenna with elongate first segment, and five following segments with long erect setae. Frons with two large foveae and a median sulcus posteriorly. Pronotum slightly cordate. Each elytron with 12 regular (and no abbreviated) striae; 4th interval with 3 foveate punctures.

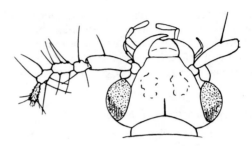

Fig. 95. Head of *Loricera pilicornis* (F.).

54. *Loricera pilicornis* (Fabricius, 1775)
  Fig. 95; pl. 4: 12.

Carabus pilicornis Fabricius, 1775, Syst. Ent.: 293.

6-8.5 mm. In general outline reminding of an *Agonum*. Black with a brass or green, rarely bluish, lustre. Mouth-parts, legs (except femora), and in part antennal bases, rufous. Elytra sometimes rufinistic. Male with 3 dilated pro-tarsal segments.

Distribution. Generally distributed in the entire area. — A truly circumpolar species; entire Europe.

Biology. Predominantly a eurytopic forest species, not preferring any particular forest community; also in the open countryside. On humid, usually muddy soil on more or less shaded ground, often at the margins of standing waters. It is a carnivorous species preying upon Collembola, insect larvae, etc. The enlarged setae of the basal segments of the antennae are used to enclose the prey during the attack (Bauer, 1982). The beetles are mainly nocturnal. *Loricera* has a well-developed ability for flight, and has been found in numbers in sea drift. It is most numerous in May-June when breeding, but is also common in the autumn when the new generation of adults emerges.

Tribe Scaritini

This and the following tribe (Broscini) are distinct within the family through the "pedunculate" body: the meso-thorax and the extreme base of elytra are strongly constricted as a "neck", upon which the scutellum is situated (Fig. 96). The front tibiae (except in *Miscodera*) are broad and spiny, adapted for digging, indicating a subterraneous mode of life.

Genus *Clivina* Latreille, 1802

Clivina Latreille, 1802, Hist. Nat. Crust. Ins. 3: 96.
  Type-species: *Tenebrio fossor* Linnaeus, 1758.

Elongate, cylindrical and somewhat flattened species (Fig. 96). Easily distinguished from *Dyschirius* through the pronotum, which is margined to base (Fig. 97), and through the strongly dilated and spiny pro- and meso-tibiae (Figs 98, 99). Upper surface without metallic lustre. Frons with a central fovea. Elytra with complete, punctate striae and a continuous row of setiferous punctures along side-margin. Meso-tibiae with a strong subapical spine (Fig. 98). Male usually without external characteristics.

This large genus has a world-wide distribution but is abundant only in warmer regions where it ecologically replaces *Dyschirius*. Our two species are subterraneous but not riparian.

Fig. 96. *Clivina fossor* (L.), length 5.5-6.5 mm. (After Victor Hansen).

Key to species of *Clivina*

1   Piceous or dark brown, first elytral interval often rufous.
Last sternite with moderately strong microreticulation .. 55. *fossor* (Linnaeus)

–   Elytra paler than forebody, yellow or reddish, usually with
a dark vitta along the suture. Last sternite dulled by strong,
granulate microsculpture .... 56. *collaris* (Herbst)

55. *Clivina fossor* (Linnaeus, 1758)
Figs 97, 99, 100, 102; pl. 4: 13.

Tenebrio fossor Linnaeus, 1758, Syst. Nat. ed. 10: 417.

5.5-6.5 mm. Piceous or dark brown (except for immaturity). First elytral interval often
rufous. All appendages pale. Certain individuals retain a yellow-brown colour trough-
out life and may then be confused with *collaris*. First pro-tarsal segment with a dentiform prominence (Fig. 99). Wings dimorphic, in short-winged specimens truncate just
outside stigma. Last abdominal sternite with moderately strong micro-reticulation. In
the male of *fossor* the median pair of the four marginal setae is more closely set than
in both sexes of *collaris*.

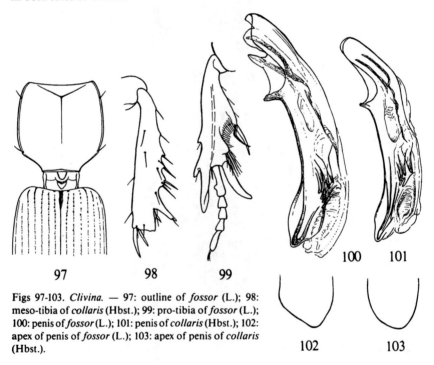

Figs 97-103. *Clivina.* — 97: outline of *fossor* (L.); 98:
meso-tibia of *collaris* (Hbst.); 99: pro-tibia of *fossor* (L.);
100: penis of *fossor* (L.); 101: penis of *collaris* (Hbst.); 102:
apex of penis of *fossor* (L.); 103: apex of penis of *collaris*
(Hbst.).

Distribution. Commonly distributed over the entire area. — Whole Europe; introduced in North America.

Biology. This is a eurytopic species, usually occurring in open countryside on rather humid ground with a more or less dense vegetation of grasses; preferably on clayish soil, never on pure sand. The diet of *C. fossor* consists of both vegetable and animal matters, for instance larvae and pupae of *Meligethes*. The species has a high flying ability and often occurs in sea drift. During daytime *C. fossor* hides in burrows in the soil or under stones. It is a spring breeder with little autumn activity.

56. *Clivina collaris* (Herbst, 1784)
Figs 98, 101, 103.

Scarites collaris Herbst, 1784, Arch. Insectengesch. 5: 141.
Buprestis contracta Fourcroy, 1785, Ent. Paris. 1: 50.

5-5.5 mm. Closely allied to *fossor* and formerly often regarded as a variety of this. Somewhat smaller and flatter, elytra shorter and with more rounded sides. Pale, forebody darker than elytra; these as a rule with a dark vitta along the suture; abdomen darker. Immaculate specimens may be distinguished from pale (immature) specimens of *fossor* by the very strong, granulate microsculpture of the last abdominal sternite, the four marginal setae of which are equidistant in both sexes. Wings full. Penis (Fig. 101) shorter with apex (Fig. 103) more rounded; also differences in details of the inner armature.

Distribution. Denmark: very rare, EJ (Vejle) and NEZ (Copenhagen area). — Sweden: found in areas in the region of Stockholm and Gothenburg, in and near gardens; recently a breeding population was found close to the seashore in western Skåne (Ålabodarna). — Not in Norway or Finland. — Europe and W. Asia.

Biology. On humus-rich soil, in Scandinavia predominantly on cultivated areas, for instance debris in gardens and greenhouses. In Denmark it has also been found in marshes and on lake shores. In C. Europe (and apparently also i Skåne) *C. collaris* is a littoral species, inhabiting eutrophic fens, river banks and salt marshes. It is most numerous in spring and early summer when breeding.

Genus *Dyschirius* Bonelli, 1810

Dyschirius Bonelli, 1810, Obs. Ent. 1: Tab.
Type-species: *Scarites thoracicus* Rossi, 1790.

Small (2-6 mm) species (Fig. 104). Body more or less cylindrical and usually with a metallic lustre (single individuals of most species may be virtually black or slightly rufinistic). Mandibles strong, arcuate (Figs 111-113); antennae short. Pronotum con-

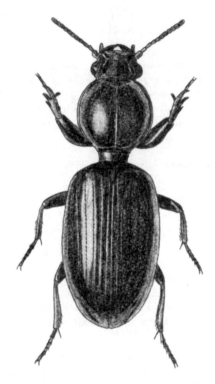

Fig. 104. *Dyschirius thoracicus* (Rossi),
length 3.5-4.7 mm. (After Victor Hansen).

vex with rounded sides and a narrow constricted base (Fig. 105). Elytra with 8 striae
or rows of punctures; 3rd interval (or stria) with 1-3 dorsal punctures. Setiferous punctures also present along side-margin of elytra and divided into two separated groups
(Figs 106, 107): (a) 1-3 subhumeral foveae (rarely lacking), each containing 2 granulae,
the posterior of which carries a seta; (b) 1-3 preapical punctures. These punctures are
taxonomically important but their examination requires good optical aids. A single
setiferous pore-puncture is present at the base of 1st stria (except in *importunus* and
laeviusculus). The male sex is distinguished externally by the somewhat broader palpal
segment.

Species of *Dyschirius* are subterranean and usually dig burrows in sterile, sandy soil.
The main prey of both imago and larva is usually staphylinid beetles of the genus
Bledius. Some species are restricted to the seashore, others to the margins of inland
waters. They catch their prey either on the surface or in the subterranean tubes of
Bledius. In most species reproduction takes place in spring or early summer. The new
generation of beetles emerges in late summer or autumn and hibernates. Larsen (1936)
described in detail the association between a number of Danish species of *Dyschirius*
and *Bledius*.

Key to species of *Dyschirius*

1　Elytra with a raised margin from shoulder to "neck". Sub-
　　humeral foveae absent 2
－　Elytra not margined inside shoulder (Fig. 105). At least
　　one subhumeral fovea (Fig. 106) 4
2(1)　Anterior margin of clypeus with a median tooth (Fig.
　　111). Two dorsal punctures in third elytral interval. 3
－　Clypeus without a median tooth. Only one dorsal punc-
　　ture in third elytral interval. (Frons rugose. Pro-tibia with
　　two small sharp teeth internally. Small and narrow) 59. *angustatus* (Ahrens)
3(2)　Entire upper surface dull from dense and strong micro-
　　sculpture. Elytral striae smooth, or almost so . . . . 58. *obscurus* (Gyllenhal)
－　At least elytra rather shiny; microsculpture clearly visible
　　only at base and apex . . . . 57. *thoracicus* (Rossi)
4(1)　Pro-tibia externally with a tubercle at base of the strong
　　apical spine (Fig. 116). Elytra with one subhumeral fovea
　　(obsolete in *neresheimeri*) 5
－　Pro-tibia externally with at least one sharp subapical
　　tooth (Fig. 117). Either 2 or 3 subhumeral foveae (as in
　　fig. 106) 9

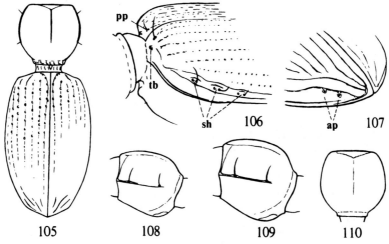

Figs 105-110. Pronotum and elytra of *Dyschirius*. — 105: outline of *luedersi* Wagn.; 106: base of elytra of *luedersi* Wagn., tb = basal tubercle of elytra, pp = basal pore-puncture and sh = sub-humeral foveae; 107: apex of elytra of *impunctipennis* Daws., ap = preapical punctures; 108: pronotum of *globosus* (Hbst.); 109 & 110: pronotum of *aeneus* (Dej.).

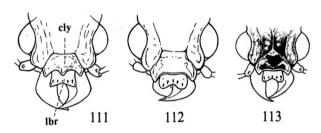

Figs 111-113. Heads of *Dyschirius*. — 111: *thoracicus* (Rossi), cl = clypeus, lbr = labrum; 112: *salinus* Schaum; 113: *luedersi* Wagn.

5(4) Elytral striae very strong, virtually impunctate; intervals
       convex throughout . . . . 63. *impunctipennis* Dawson
  –   Elytral striae moderately impressed, evidently punctate at
       least anteriorly; intervals flat or convex near the suture 6
6(5) Terminal spine of pro-tibia strongly arcuate, much longer
       than terminal spur (see fig. 115). Suture of elytra impres-
       sed at base. Large species . . . . 64. *chalceus* Erichson
  –   Terminal spine of pro-tibia moderately arcuate, hardly
       longer than terminal spur. Suture little impressed 7
7(6) Elytral striae very fine, disappearing before apex. Nar-
       row with parallel-sided elytra. Less than 5 mm . . . . 62. *politus* (Dejean)
  –   Elytral striae well impressed, complete though weaker to-
       wards apex. Elytra not quite parallel-sided. Usually more
       than 5 mm 8
8(7) Elytra almost constantly with 3 dorsal punctures; sub-
       humeral fovea evident . . . . 60. *nitidus* (Dejean)
  –   Elytra with 2 dorsal punctures; subhumeral fovea obso-
       lete . . . . 61. *neresheimeri* Wagner
9(4) Raised lateral bead of pronotum reaching posterior setife-
       rous puncture (Fig. 109). Wings full (except in *nigricornis*) 10
  –   Marginal bead of pronotum developed anteriorly only
       (Fig. 108). Wings almost constantly rudimentary 15
10(9) Posterior limit of clypeus sharply angulate (Fig. 113) 11
  –   Clypeus delimited towards frons by a straight or ob-
       tusely angulate line 13
11(10) Sloping base of elytra with a small tubercle (tb in Fig.
       106). Frons with a median ridge forming a continuation
       of the clypeal angle (Fig. 113) . . . . 67. *luedersi* Wagner
  –   Elytral base without such tubercle. Clypeal angle not pro-
       longed upon frons 12

96

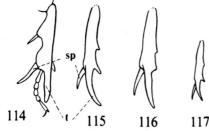

Figs 114-117. Left pro-tibia of *Dyschirius*. — 114: *thoracicus* (Rossi), front view, sp = spur, t = terminal spine; 115: same species, external view; 116: *politus* (Dej.); 117: *globosus* (Hbst.).

12(11) Three subhumeral foveae, two preapical punctures (as in figs 106, 107). Elytral striae fine apically but reaching apex .... 66. *aeneus* (Dejean)
- Two subhumeral foveae, one preapical puncture. Elytral striae obsolescent towards apex .... 68. *septentrionum* Munster
13(10) Antennae and legs entirely black. One preapical puncture. Posterior limit of clypeus arcuate .... 69. *nigricornis* Motschulsky
- Antennal base and at least part of tibiae pale. Two preapical punctures. Posterior limit of clypeus straight. Base of elytra with a small tubercle (tb in fig. 106) 14
14(13) Three subhumeral foveae. Elytral striae strongly punctate .... 65. *salinus* Schaum
- Two subhumeral foveae. Elytral striae finely punctate .... 70. *intermedius* Putzeys
15(9) Elytra with a large pore-puncture at base of first stria .... 72. *globosus* (Herbst)
- Elytra without basal pore-puncture .... 71. *laeviusculus* Putzeys

The first three species (nos. 57-59) have a raised margin from shoulder to neck and a subhumeral fovea is lacking.

57. *Dyschirius thoracicus* (Rossi, 1790)
Figs 104, 111, 114, 115.

Scarites thoracicus Rossi, 1790, Fauna Etrusca 1: 227.
Dyschirius arenosus Stephens, 1827, Ill. Brit. Ins. Mand. 1: 42.

3.5-4.7 mm. Distinguished (together with *obscurus*) by the median tooth of the clypeus (Fig. 111). Short and convex, elytra with rounded sides. Usually with a strong brassy lustre, sometimes bluish, rarely unmetallic black or piceous; mouth-parts, base of antennae and at least tibiae pale. Subhumeral fovea absent; two dorsal and one preapical punctures present. Elytral striae clearly punctate, at least anteriorly. Elytral microsculpture obsolete on central parts. Male with somewhat narrower elytra and often a stronger metallic lustre than female.

Distribution. In Denmark generally distributed and rather common. — Sweden: rather distributed from Sk. to Hls., also around the northern coast of the Bothnian Bay. — Norway: along the southern coast but also one record from Nordland. — Finland: generally distributed north to the Polar Circle; also a few finds further north (lake Enare); also Vib, Kr and Lr in the USSR. — West coast of Europe; northern parts of E. Europe and further east; N.Africa and W.Asia.

Biology. On sterile sandy shores of fresh and salt water, preferably on moist, fine sand, but also on rather coarse and clay-mixed sand. On the seashore usually in colonies of *Bledius fergussoni* Joy (= *arenarius* Payk.), together with *D. obscurus* and *Bembidion pallidipenne;* on lake shores and river banks often with *Bledius talpa* Gyll., but also in association with other species of *Bledius*. It is most numerous in spring and late summer.

58. *Dyschirius obscurus* (Gyllenhal, 1827)

Clivina obscura Gyllenhal, 1827, Ins. Suec. 4: 456.

3.5-4.6 mm. Closely allied to *thoracicus* and differing only in the following points: elytra shorter with more rounded sides; entire upper surface dull from dense and strong microsculpture; legs and base of antennae usually darker; elytral striae smooth, or almost so, and deeper, notably at apex. Male slightly narrower.

Distribution. Denmark: rare and local along the coasts; also two records from inland localities in SZ. — Sweden: rare and local on the south coast from Boh. to Gtl.; also at the Bothnian Bay in Ång., Vb. and Nb. — Norway: only Ry (Jæren). — Finland: on the southern and north-western coasts; also a few inland localities; Vib and Kr (at lake Ladoga). — Mostly on the Baltic and west coasts of Europe, south to N.France; scattered in the interior parts, east to the Caspian Sea.

Biology. More stenotopic than *D. thoracicus,* being dependent on finer and more humid sand. It is almost restricted to the seashore, usually in shifting sand habitats, often together with *D. impunctipennis* and *Bembidion pallidipenne* in colonies of *Bledius fergussoni* Joy (= *arenarius* Payk.). Rarely in similar habitats on lake shores (e.g. Lake Ladoga, USSR), and in sand pits (Denmark). Predominantly in spring and late summer.

59. *Dyschirius angustatus* (Ahrens, 1830)

Clivina angustata Ahrens, 1830, Ent. Arch. 2 (2): 60.

3-3.4 mm. Easily distinguished by the extremely narrow body (still narrower than in *politus*), the rugose frons, and the margined base of elytra (as in the two preceding species).
Piceous, upper surface with a bronzy hue, underside of pronotum, elytral

epipleura, frons anterior, antennal base, mouth-parts, and main parts of legs, rufous brown. Frons dull from coarse rugosity and punctuation. Pro-tibiae with small but sharp teeth externally. The species is unique in having only one dorsal puncture on the elytra. No subhumeral foveae, one preapical puncture.

Distribution. In Denmark very local and rare: WJ (4 localities in the Esbjerg-area), SJ + EJ (8 localities), NWJ (Bulbjerg), F (Ristinge and Ærø), LFM (Møn), NEZ (Lynæs and Tisvilde). — Sweden: very rare and extremely local, recorded from 6 localities: Sk. (Ålabodarna and Ven), Öl. (Resmo), Vrm. (Höje), Ång. (Bjurholm at river Öreälven), Lu. Lpm. (Messaure). — Norway: scattered localities between 66°N and 70°N. — In East Fennoscandia only a few records from the far north. — A European species, but not in the south, and everywhere very local.

Biology. On humus-mixed, sandy and often clayey slopes near water. In Denmark and southern Sweden (Sk.) preferably on sea-slopes. Elsewhere in Scandinavia mainly on steep river banks, rarely on lake shores (N.Finland). The species is apparently always associated with *Bledius*, e.g. *B. nanus* Er. *filipes* Sharp, *erraticus* Er. and *longulus* Er.. Mostly in June-July.

The species nos. 60-72 are without a raised margin on elytra inside shoulder. One, 2 or 3 subhumeral foveae are present. The first five species have pro-tibiae only tuberculate externally, and one subhumeral fovea.

60. *Dyschirius nitidus* (Dejean, 1825)

Clivina nitida Dejean, 1825, Spec. Gén. Col. 1: 421.

4.5-5.5 mm. Reminding of *chalceus* but pronotum and elytra more convex, with somewhat more rounded sides, and elytra without sutural impression at base. Striae evident to apex, more strongly punctate than in *politus*. Pro-tibiae only tuberculate externally. One subhumeral fovea; 3 dorsal and 2 preapical punctures.

Distribution. Not found in Denmark, Sweden or Norway. — In Finland a few old records in N and Sa, but no records from the 20th century. — In the USSR abundant in some localities around lake Ladoga. — W., C. and S.Europe; Siberia.

Biology. In the Ladoga-area it is confined to steep river banks, occurring on humus-mixed fine sand with sparse vegetation. Elsewhere in Europe also shores of lakes and ponds, and on the sea-coast. Always in association with *Bledius*, e.g. *B. opacus* Block, *subterraneus* Er. and *praetermissus* Will. (= *atricapillus* Germ.).

61. *Dyshirius neresheimeri* Wagner, 1915

Dyschirius Neresheimeri Wagner, 1915, Ent. Mitt. 4: 241.

5-5.6 mm. Very closely related to *nitidus* but somewhat broader. Elytra at base dulled by microsculpture, with only 2 dorsal punctures; their striae finer.

Distribution. Not found in Denmark, Norway or Finland. — Sweden: Upl., Runmarö, 1 specimen, no doubt arrived by accidental drift (G. Hoffstein, RM). — C.Europe: Germany, Holland, Poland, W. Russia. Nearest population is found in Latvia.

Biology. Little is known concerning the biology and ecology of this very local and rare species. In C.Europe it has been found on sand and clay on the border of standing fresh water, for instance in clay pits.

62. *Dyschirius politus* (Dejean, 1825)
Fig. 116; pl. 4: 14.

Clivina polita Dejean, 1825, Spec. Gén. Col. 1: 422.

4-4.9 mm. Almost as slender as *angustatus,* but much larger and with a smooth frons. Brassy, rarely with a bluish hue, elytra often rufinistic; legs quite pale or with femora somewhat infuscated, tibiae palest. Elytral striae very fine, faintly punctate, obsolete towards apex, extreme sloping base micro-reticulate. One subhumeral fovea; 2 dorsal and 2 preapical punctures.

Distribution. Denmark: very rare and sporadic, recorded with decreasing frequency. — Sweden: local and uncommon, recorded from most districts except several of the Lappmarks. — Norway: in some southern districts. — Finland: north to 67°N, rather sporadic; also in the southern areas of the Soviet part of East Fennoscandia. — This species and *nigricornis* are the only true circumpolar species of *Dyschirius;* Europe, except the south; Siberia, North America.

Biology. At the margins of standing or running waters, on very fine sand with a muddy coating of the surface and usually sparsely vegetated by *Carex, Equisetum,* etc. It prefers somewhat drier places than *D. thoracicus* and is often found at some distance from the water edge. Especially abundant on the banks of fair-sized rivers, associated with *Bledius fuscipes* Rye, *opacus* Block and *longulus* Er. Also in sand pits, on sea-slopes, and occasionally on the sea-coast in colonies of *B. fergussoni* Joy (= *arenarius* Payk.). Mainly in June-July.

63. *Dyschirius impunctipennis* Dawson, 1854
Fig. 107.

Dyschirius impunctipennis Dawson, 1854, Geod. Brit.: 29.

4.5-5.2 mm. Very characteristic through the strong, complete, virtually impunctate elytral striae and the convex intervals. Body almost as narrow as in *politus,* with parallel-sided elytra. Mandibles conspicuously strong and arcuate. Labrum deeply sinuate. Black with a quite faint metallic reflection, elytra not seldom rufinistic. Elytral punctures as in the following species.

Distribution. In Denmark very rare and scattered on the coast, SJ (Rømø), WJ +

NWJ + NEJ (several localities from Fanø to Skagen), EJ (Anholt), LFM (Gedser, Bøtø, Høje Møn), NWZ (Nykøbing) and B (Dueodde, Nexø). — Sweden: very rare; local records from Sk., Hall., Öl., Gtl., G. Sand., Ög., Vg., Dls. After 1950 only very few records. — Norway: Ry (Jæren). — Finland: a few localities at the Gulf of Bothnia from Al to ObS.; also Vib and at lake Ladoga. — Europe, along the west coast and locally inland.

Biology. Almost confined to moist, sterile, fine sand on the sea-coast, usually in shifting sand habitats, occurring in colonies of *Bledius fergussoni* Joy (= *arenarius* Payk.), often together with *D. obscurus* and *Bembidion pallidipenne*. Rarely on inland localities, in our area on the banks of lake Ladoga (USSR). It is most numerous in early summer.

64. *Dyschirius chalceus* Erichson, 1837

Dyschirius chalceus Erichson, 1837, Käf. Mark Brandenburg 1 (1): 38.

Our largest species: 5.2-6 mm. Rather broad but with conspicuously parallel-sided elytra. The lustre is brassy, but the elytra are often brownish translucent as in *politus*. Antennal base, mouth-parts and legs more or less pale (as usual in this group). The impression on the base of the suture is diagnostic. One subhumeral fovea; 2 dorsal (none before middle) and 2 preapical punctures.

Distribution. Denmark: very rare; WJ (Fanø, Esbjerg, Skallingen), LFM (Vålse Vig, 1887), NEZ (København, Amager). — Sweden: very rare and local in the south, Sk., Hall., Öl., Gtl. — Not in the rest of Fennoscandia. — Coasts of W. and S.Europe, also scattered in saline localities in the interior of Europe; W.Asia.

Biology. A halobiontic species, inhabiting salt marshes on the sea-coast. It occurs on moist clay or silt with sandy subsoil, in habitats often flooded by the tide. The salt content of the soil water is usually above 2% and may rise to almost 10% (Larsen, 1936). In C.Europe also in saline inland localities. The beetles dig deep, vertical shafts into the ground. They live in association with large species of *Bledius*, e.g. *B. furcatus* Ol., *spectabilis* Kraatz, *diota* Schiö. and *tricornis* Hbst., often together with *D. salinus*. Most numerous in May-June, single specimens are seen in the autumn.

The species nos. 65-72 all have at least one sharp subapical tooth externally on protibia, and 2 or 3 subhumeral foveae.

65. *Dyschirius salinus* Schaum, 1843
    Fig. 112.

Dyschirius salinus Schaum, 1843, Z. Ent. (Germar) 4: 180.

3.6-4.5 mm, and thus larger than the other members of this group, differing from these by the straight posterior limit of clypeus (Fig. 112). Somewhat similar to *thoracicus* but with punctuation of elytral striae much stronger. Black with a faint metallic lustre, elytra sometimes rufinistic. A small tubercle (Fig. 106) is present on the basal slope of elytra (as in *luedersi*). Three subhumeral foveae; 3 dorsal and 2 preapical punctures.

Distribution. In Denmark rare, all localities situated at the coasts. — Sweden: rather common along the west coast; also found in Sm., Öl. and Gtl. — Norway: a few localities in the south-eastern part. — In E. Fennoscandia only in the south-west part of Finland (abundant in Al, a few records in Ab and N). — In Europe along almost all the coasts; also isolated in the inland; N.Africa, W.Asia.

Biology. A halobiontic species, in our area confined to salt marshes on the sea-coast. It lives on moist sand mixed with clay or silt, both on bare ground and in sparse vegetation of for instance *Plantago maritima*. The species is often found together with *D. luedersi* and is usually associated with, but apparently not dependent on *Bledius*, e.g. *B. diota* Schiö., *furcatus* Ol. and *tricornis* Hbst. In C.Europe *D. salinus* is also known from saline inland localities. It is most numerous in May-July.

66. **Dyschirius aeneus** (Dejean, 1825)
    Figs 109, 110, 118.

Clivina aenea Dejean, 1825, Spec. Gén. Col. 1: 423.

3.1-3.6 mm. Closely related to the two following species but smaller and elytra shorter. The angulate posterior limit of clypeus is not prolonged into a ridge. Pronotum (Fig. 110) with its greatest width behind middle. Sides of elytra more parallel, their base with the setiferous pore-puncture in a higher position and the tubercle is absent. Similarly coloured, except that the antennal base is paler, reddish. Subhumeral fovea, etc., as in *salinus*. Penis as in fig. 118.

Distribution. Denmark: rare in Jutland (a few localities in the east); local on the islands; not B. — Sweden: rare, scattered localities north to Vrm. and Dlr. — Norway: AK, Bø and TEy. — Finland: a few localities in Al, Ab and N; once found in Kb: Tohmajärvi (1984). — Most of Europe and parts of Asia.

Biology. An inhabitant of the border of standing and slow-running fresh water. It occurs on wet, muddy-clayish soil, on bare spots in otherwise rather dense vegetation of *Carex* spp., *Equisetum hiemale, Alisma plantago-aquatica,* etc. It is rarely found with *Bledius,* but is often associated with *Heterocerus* (Heteroceridae) and *Carpelimus* (= *Trogophloeus;* Staphylinidae). These may be the main prey. Most numerous in June.

67. **Dyschirius luedersi** Wagner, 1915
    Figs 105, 106, 113, 120.

Dyschirius Lüdersi Wagner, 1915, Ent. Mitt. 4: 304.
Dyschirius unicolor auctt.

3.4-4.1 mm. Was earlier confused with *aeneus* from which it is best separated by the median keel of frons (Fig. 113) and the small tubercle on the elytral base (Fig. 106). Black, usually with a brassy lustre, exceptionally without or with a bluish lustre. Base of antennae, mouth-parts, and legs piceous. Pronotum widest at middle, sides equally rounded before and behind the middle. Elytra more widened behind shoulder than in *aeneus*. Armature of pro-tibiae less developed than in *salinus*. Subhumeral foveae etc. as in *globosus*. Penis as in fig. 120.

Distribution. Denmark: generally distributed and rather common. — Sweden: distributed and rather common in some districts in the south, north to Upl.; isolated at the Bothnian Bay (single localities in Ång. and Nb.). — Norway: in the extreme southeastern districts. — In E. Fennoscandia recorded north to Ks and southern Lr. — Almost entire Europe except the south. Asia, at least to W.Siberia.

Biology. Mostly on firm clayey soil, usually at the border of standing water, but also on the banks of slow-running rivers. It is especially predominant in salt marshes on the seashore; in C. Europe also in saline inland localities. The beetle is usually found on bare spots in areas otherwise rather densely covered with *Carex* spp., saltmarsh plants etc. Species of *Bledius* are only occasionally found together with *D. luedersi,* the prey of which is rather species of *Heterocerus* and *Carpelimus* (=

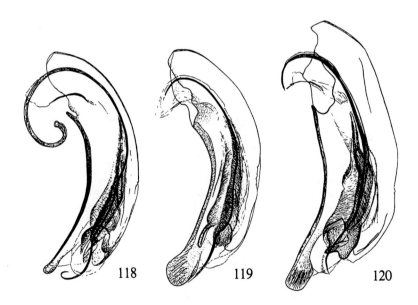

Figs 118-120. Penis of *Dyschirius*. — 118: *aeneus* (Dej.); 119: *septentrionum* Munst.; 120: *luedersi* Wagn.

Trogophloeus). Regularly occurring in drift on the sea-coast. The species is most numerous in spring when breeding, and in late summer when newly emerged adults occur.

68. *Dyschirius septentrionum* Munster, 1923
Fig. 119.

Dyschirius septentrionum Munster, 1923, Norsk Ent. Tidsskr. 1: 248.

2.8-3.6 mm. In general proportions most like *aeneus*. Can usually be separated from the two preceding species by having only 2 subhumeral foveae and one preapical puncture, and by the apically evanescent striae. However, a third subhumeral fovea may be suggested and a second preapical puncture be present, but such individuals are exceptional as are specimens with more well developed striae. The base of antennae is either piceous, as in *luedersi,* or clear red, as in most *aeneus*.

The species may always be separated from *luedersi* on the structure of frons and elytral base (cf. figs. 106, 113); from *aeneus* by the form of the pronotum (more like that of *luedersi*) and the more rounded elytral sides. Colours as in the two species compared with. Penis as in fig. 119.

Distribution. Not in Denmark. — In Sweden distributed in the north, south to Vrm., especially frequent along the Bothnian Bay. — Norway: from 60°N to the extreme north. — In Finland generally distributed north of 62°N; also Vib, Kr and Lr. — Palaearctic region, from Fennoscandia east to Kanin, and in the Ob and Lena river districts.

Biology. Especially typical of river banks, occurring on clayey or muddy sites with vegetation of *Juncus, Carex, Equisetum,* etc., often in thin growth of *Salix* and *Alnus*. Less common on the borders of standing waters, e.g. small ponds, in clay pits and on the seashore. It is often found in company with *Bembidion schuppelii* and *semipunctatum* (Andersen, 1982). Several species of *Bledius* are usually present in the habitat of *D. septentrionum,* e.g. *B. fuscipes* Rye, *arcticus* J. Sahlb. and *longulus* Er., but *D. septentrionum* seems not to have any obligatory dependence on these. In the Scandinavian mountains the species is found up to the birch region. The adults occur throughout the summer; newly emerged adults appear in July and August and then hibernate. Development may last two years under the climatic conditions of northern Scandinavia (Lindroth 1945).

69. *Dyschirius nigricornis* Motschulsky, 1844

Dyschirius nigricornis Motschulsky, 1844, Ins. Sibér.: 82.
Dyschirius Helleni J. Müller, 1922, Koleopt. Rdsch. 10: 45.
Dyschirius norvegicus Munster, 1923, Norsk Ent. Tidsskr. 1: 249.

2.6-3 mm. A small and narrow species with entirely black appendages. Upper surface black with a faint brass lustre. Posterior limit of clypeus obtusely angulate or horseshoe-shaped. Elytra broader behind middle; striae usually reaching apex, often somewhat irregular but strongly punctate basally. Two subhumeral foveae; 3 dorsal and 1 preapical punctures. Wings rudimentary, narrower and shorter than one elytron.

Distribution. Not in Denmark. — Sweden: rare, in Lapland several localities south to 65°N and a single locality in Jmt. — Norway: On (Dovre), TRy and Fi. — Finland: north of the Polar Circle; also Lr. — Circumpolar in high latitudes, east to Labrador.

Biology. A stenotopic species, strictly confined to bogs in the upper conifer and the birch region of northernmost Fennoscandia. According to Krogerus (1960) it shows a strong preference for low temperatures (about 10°C) and acid soil (pH 3.8-4.2). *D. nigricornis* lives among moss and dead leaves in more or less dry places with dwarf shrub vegetation of *Betula nana, Vaccinium uliginosum, Rubus chamaemorus* and *Empetrum,* and with tufts of *Sphagnum fuscum.* It is most easily collected by sifting. The species is not associated with *Bledius.* Adults occur from June to September and are most numerous in July. They, and probably also larvae, hibernate. The development may last more than one year (Lindroth 1945).

70. *Dyschirius intermedius* Putzeys, 1846

Dyschirius intermedius Putzeys, 1846, Mem. Soc. R. Sci. Liège 2: 550.
Dyschirius sylvaticus Thomson, 1859, Skand. Col. 1: 14.

3-3.5 mm. Most like *aeneus* but separated from this as well as from the two other related species by the straight posterior limit of clypeus. Elytra narrower and more stretched than in *aeneus* and with shallower striae, their sloping base with a small tubercle as in *luedersi* (Fig. 106). Two subhumeral foveae; 3 dorsal and 2 preapical punctures. Antennae and legs markedly pale.

Distribution. Denmark: very rare and local. - Sweden: Sk., Bl., Hall., Gtl., Vg.; rare and local. — Not in Norway or Finland. — USSR: river Svir area. — Only in Europe, east to S.Russia.

Biology. This southern species usually lives at the border of standing or running waters, especially where the banks are steep, for instance on sea-slopes. It prefers fine clayey sand with a muddy surface and usually sparsely vegetated. The species occurs in association with *Bledius nanus* Er., *fuscipes* Rye, *longulus* Er. and *erraticus* Er., and sometimes together with *D. politus* and *angustatus.* It is most numerous in June.

71. *Dyschirius laeviusculus* Putzeys, 1846

Dyschirius laeviusculus Putzeys, 1846, Mem. Soc. R. Sci. Liège 2: 547.

3-3.5 mm. Black with a metallic lustre. Very similar to *globosus* but (as in *importunus*) without basal pore-puncture on elytra. Colour as in *globosus,* that is with at least pro-femora dark. Frons with a distinct tubercle. Elytra narrower and larger, striae virtually disappearing apically. Wings full. Two or three dorsal punctures. Otherwise as in *globosus.*

Distribution. Not in Denmark, Norway og Finland. — Sweden: Sk. — E. & C.Europe south to Yugoslavia and the Pyrenees; present in the Baltic States.

Biology. This species was found in 1974 at Ålabodarna, Sk. (Lundberg, 1980). It lives on river banks and at the border of ponds. On the Swedish locality it was taken close to the sea in the burrows of *Bledius praetermissus* Will. and *B.nanus* Er.

72. *Dyschirius globosus* (Herbst, 1784)
Figs 108, 117.

Carabus globosus Herbst, 1784, Arch. Insectengesch. 5: 142.
Scarites gibbus Fabricius, 1792, Ent. Syst. 1: 96.

2.2-3 mm. Our smallest and shortest species, with sides of pronotum and elytra strongly rounded. Black or piceous with a faint metallic lustre. Base of antennae, mouthparts and legs (except pro-femora) pale. Rufinistic specimens, especially such with a pale pronotum, are not uncommon. Elytral striae strongly punctate anteriorly, more or less evanescent towards apex. Three subhumeral foveae; 3 dorsal and 3 preapical punctures. Constantly short-winged in our area, but macropterous specimens have been found on the continent.

Distribution. Generally distributed and very common in all four countries; one of the commonest carabids at all in our area. — Entire Europe, N.Africa, Siberia.

Biology. A very eurytopic species, inhabiting almost every kind of moderately humid soil, e.g. clay, sand and peaty soil with rather sparse vegetation. It occurs on the border of lakes and rivers as well as far from water, for instance on heaths; also on arable land and in saline habitats on the sea-coast. The species is not associated with *Bledius,* but has been observed preying upon species of *Carpelimus (= Trogophloeus).* The adults become active in early spring and are found almost throughout the year.

Editors note. *Dyschirius importunus* (Schaum, 1857) has been recorded from the southern Soviet part of East Fennoscandia. The reported specimen(s) is no longer available and the record must be strongly doubted, as the true *importunus* is a strict Mediterranean species.

Tribe Broscini

For diagnostic characters see under Scaritini. The tribe is abundantly represented on the southern hemisphere, particularly in Australia and New Zealand.

Genus *Broscus* Panzer, 1813

Broscus Panzer, 1813, Index Ent.: 62.
  Type-species: *Carabus cephalotes* Linnaeus, 1758.

One of our largest and very characteristic carabids (Fig. 121). Head almost as wide as pronotum, with enormous mandibles. Pronotum with two marginal setae. Elytra

Fig. 121. *Broscus cephalotes* (L.), length 16-23 mm. (After Victor Hansen).

transversely microsculptured, with rows of very fine punctures. Front legs broad, notably the tibiae, adapted for digging. Wings full but normally not functioning. Male with three dilated pro-tarsal segments.

73. *Broscus cephalotes* (Linnaeus, 1758)
Fig. 121; pl. 3: 3.

Carabus cephalotes Linnaeus, 1758, Syst. Nat. ed. 10: 414.

16-23 mm. Black, without any trace of metallic hue. Palpi, antennae and tarsi piceous.

Distribution. In Denmark generally distributed and common. — Sweden: north to Dlr. and Gstr. (c. 61°N), common especially in the southern parts. — Norway: several south-eastern districts and along the south coast to Ry (Jæren); isolated in the Trondheim area (probably late immigration). — In Finland plus Vib and Kr of the USSR north to 64°N. — Entire Europe except in the north and south; east to W.Siberia.

Biology. A heat-preferent, xerophilous species, occurring on open, often quite barren, dry soil. It is most predominant on sandy soil, for instance sandy grassland, coastal dunes and lake shores; also on cultivated land, in particular in root crop fields. In the more continental climate of C.Europe the species also occurs on clay-soil. The beetle digs deep burrows in the soil, often under pieces of dead wood, and stays here during the daytime; it is active at night. *Broscus* is a ravenous predator on all kinds of insects and other arthropods, on sandy beaches for instance on the amphipod *Talitrus saltator*. Reproduction takes place in late summer and autumn, eggs are being deposited in deep shafts in the soil. Adult beetles occur from spring to late autumn, mostly in June-August.

Genus *Miscodera* Eschscholtz, 1830

Miscodera Eschscholtz, 1830, Bull. Soc. Nat. Mosc. 2: 63.
Type-species: *Scarites arcticus* Paykull, 1798.

Comprises a single very convex species with a narrow head (Fig. 122). Pronotum almost spherical, with only one (the anterior) marginal seta. Each elytron with four basal foveae, the striae replaced by rows of punctures disappearing towards apex. Antennae short, with terminal segments rounded. Pro-tibiae only slightly modified, not adapted for digging. Wings full and functional. Pro-tarsi of male with 3, meso-tarsi with 2, dilated segments.

74. *Miscodera arctica* (Paykull, 1798)
Fig. 122; pl. 4: 15.

Scarites arcticus Paykull, 1798, Fauna Suec. Ins. 1: 85.

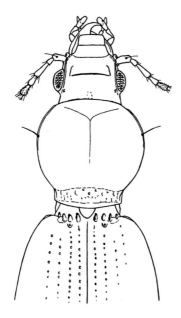

Fig. 122. *Miscodera arctica* (Payk.).

6.5-8 mm. Piceous to almost black, upper surface usually brassy, rarely bluish or somewhat rufinistic. All appendages rufo-piceous.

Distribution. Denmark: very rare, EJ (Mols 1966-80), NEJ (Blokhus 1925, Ålbæk 1873, Læsø 1979, several fir plantations on the Skaw 1980), B (Hammeren 1905, Dueodde 1979). — Sweden: in most districts but most abundant in the north. — Norway: generally distributed except along the west coast. — Finland and Soviet part of E.Fennoscandia: entire area but scattered. — A circumpolar species. The sparse occurrences in S.Sweden, Denmark and N.Germany have relict character.

Biology. A decidedly northern species, mainly occurring from the upper conifer to the alpine region, but also on the tundra (Kola). Found on moderately dry, fine sand, often mixed with gravel, usually in open country, but also in thin pine forest (Denmark). It is most predominant in *Calluna* and *Empetrum* vegetation, but also inhabits alpine grass heaths and poor high mountain ground. The species is regularly found together with *Byrrhus* and *Cytilus,* the larvae of which it may prey upon. *Cymindis vaporariorum, Amara quenseli* and, in the north, *Bembidion grapii* are often associated with *Miscodera.* In northern Sweden newly emerged imagines have been found in July-August, indicating that the species here hibernates as imago. Autumn reproduction and larval hibernation is more likely to occur in southern Scandinavia, because females in Denmark are found in mid-July at the beginning of the breeding period.

Tribe Patrobini

Genus *Patrobus* Dejean, 1821

Patrobus Dejean, 1821, Cat. Coll. Col. B. Dejean: 10.
Type-species: *Carabus atrorufus* Ström, 1768.

Without metallic lustre, in general habitus (Fig. 123) somewhat similar to a small *Pterostichus,* and most easily recognized by the well delimited, constricted neck. Frontal furrows deep, somewhat converging anteriorly (Figs 124, 125). Pronotum cordiform with sharp, about right hind-angles and a single, deep basal fovea, delimited externally by a carina. Elytral base not margined inside shoulder, third interval with 3-4 dorsal punctures; striae complete, punctate. Male with 2 strongly dilated pro-tarsal segments. Parameres almost symmetric, with two or more long setae at tip.

The distribution is holarctic and mainly northern. The species are carnivorous beetles.

Key to species of *Patrobus*

1 Wings fully developed (in repose with reflexed apex). Pronotum with anterior transverse impression deep, anterior margin therefore appearing elevated 2
- Wings rudimentary, reduced into a narrow scale shorter than half elytron. Anterior transverse depression of pronotum shallow 3
2(1) Slender-bodied, with parallel-sided elytra. Maxillary palpi (Fig. 127) rather thick .... 75. *septentrionis* Dejean
- Elytra conspicuously broader. Maxillary palpi (Fig. 126) more slender .... 76. *australis* J. Sahlb.
3(1) Area between frontal furrows and side-margin widening anteriorly (Fig. 124). Third antennal segment longer than first .... 78. *atrorufus* (Strøm)
- Area between frontal furrows and side-margin equally wide (Fig. 125). First and third antennal segments equally long................
.... 77. *assimilis* Chaudoir

75. *Patrobus septentrionis* Dejean, 1828
    Figs 123, 127; pl. 4: 16.

Patrobus septentrionis Dejean, 1828, Spec. Gén. Col. 1: 29.
Harpalus picipennis Zetterstedt, 1828, Fauna Ins. Lapp. 1: 32.
Patrobus hyperboreus Dejean, 1828, Spec. Gén. Col. 1: 30.
Patrobus rubripennis Thomson, 1857, Skand. Col. 1: 26.

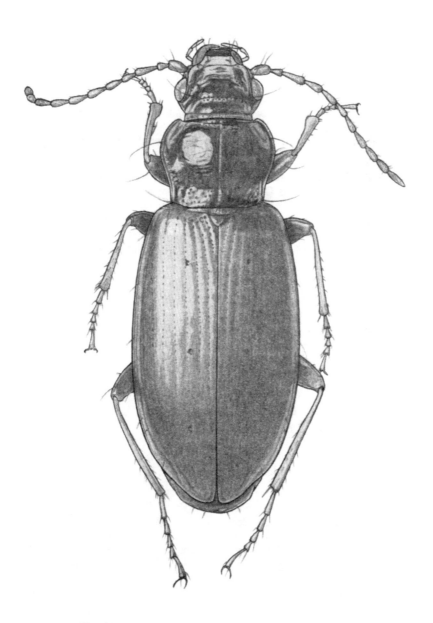

Fig. 123. *Patrobus septentrionis* Dej., length 7.4-10 mm.

7.4-10 mm. Slenderer than the following species, with shorter and laterally less sinuate pronotum, and longer, more parallel-sided elytra. Black or piceous, appendages paler, elytra often bright rufinistic and then sometimes darker along the suture. Frons intermediate between that of *assimilis* and *atrorufus* (Figs 124, 125). Elytral striae comparatively weak. Wings always full.

Distribution. Not in Denmark. — In Sweden generally distributed from northern Dlr. to T. Lpm.; rather common. — Norway: generally distributed in the north; in the south mainly in elevated areas. — Also in the northern parts of E.Fennoscandia. — Circumpolar at high latitudes; also in the Alps and Great Britain.

Biology. This northern species is a common ground-beetle in the Scandinavian mountains above the timber limit. It occurs in wet habitats like marshes and meadows as well as on not too dry heaths. Often found close to snow drifts and is reported hunting frosen insects on the snow. Also occurring in the upper conifer region, where it lives in moist places with rich vegetation of *Carex* etc. at the margins of large rivers, often together with *Pelophila borealis*. The life cycle is biennial, at least in the alpine region. Adults occur in May-September, newly emerged beetles in June-August.

76. *Patrobus australis* J. Sahlberg, 1875
Fig. 126.

Patrobus septentrionis var. *australis* J. Sahlberg, 1875, Notis. Sällsk. Faun. Fl. Fennica Förh. 14: 91.
Patrobus septentrionis relictus Neresheimer & Wagner, 1927, *in* Wagner, Coleop. Centralbl. 2: 91.

This species was earlier regarded a southern form of *septentrionis*, but is here treated as a good species. It is flatter than *septentrionis*, and the elytra are usually conspicuously broader. Also the explanate margin of pronotum is broader. Maxillary palpi (Fig. 126) more slender. The colour is never quite black, but truly rufinistic specimens do not occur.

Distribution. Denmark: very rare in Jutland (EJ: Århus and Samsø); more distributed on the islands and B. — Sweden: very rare, only recorded from a few localities in Skåne (Uppåkra, Häckeberga, Vomb, Hallands Väderö, Sandhammaren and Skanör). Recorded only after 1950. The finds from the three last-mentioned localities are accidental, caused by wind transportation. - Not in Norway. — Finland: not common, in several districts south of 62-63°N; also Vib and southern Kr in the USSR. — Total distribution imperfectly known, but includes N.Germany.

Biology. Very hygrophilous: on the muddy shores of stagnant waters, e.g. ponds, pools, clay pits, where occurring among decaying litter. Notably in forests, but also in shaded sites in the open countryside. The adult beetles occur in May-September; they probably reproduce in the autumn.

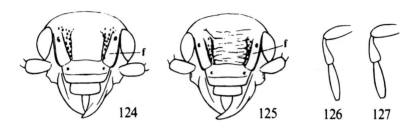

Figs 124, 125. Heads of *Patrobus*. 124: *atrorufus* (Strøm); 125: *assimilis* Chaud., f = elevated frontal field.
Figs 126, 127. Maxillary palpi of *Patrobus*. — 126: *australis* J. Sahlb.; 127: *septentrionis* Dej.

77. *Patrobus assimilis* Chaudoir, 1844
Fig. 125.

Patrobus assimilis Chaudoir, 1844, Bull. Soc. Nat. Mosc. 17: 441.
Patrobus clavipes Thomson, 1859, Skand. Col. 1: 214.

8.8-9 mm. Narrower and more convex than *atrorufus*. Also darker, upper surface almost black, suture not paler. Antennae more moniliform (with rounded segments). Head (Fig. 125) and pronotum transversely wrinkled. Pronotum narrower, more convex, basal foveae smaller but deep. Elytra with rounded shoulders and striae more coarsely punctate anteriorly. Legs, especially femora, shorter. Wings constantly rudimentary.

Distribution. Denmark: very rare, sparsely distributed in WJ, EJ, NEJ and NEZ. — Generally distributed in the whole of Fennoscandia and recorded from practically all districts. — N. and C.Europe, NW. Siberia. A boreomontan species.

Biology. In northern Fennoscandia rather eurytopic, occurring in open country as well as in forests on fairly dry ground, e.g. *Calluna* and *Empetrum* heaths, but also in wet habitats such as swamps and lake shores. It is common in the alpine region, notably on grass meadows; also often on snow-beds. In southern Scandinavia the species is exclusively found in damp localities, predominantly peaty woods with birch and alder and a rich vegetation dominated by *Sphagnum, Carex*, etc.; often in association with *Trechus rivularis*. It is an autumn breeder. In northern and alpine regions development lasts two years.

113

78. *Patrobus atrorufus* (Ström, 1768)
Fig. 124.

Carabus atrorufus Ström, 1768, Skr. K. Norske Vidensk. Selsk. 4: 331.
Carabus excavatus Paykull, 1790, Mon. Car. Suec.: 38.

7.4-10 mm. Reddish brown to piceous; appendages and often also elytral suture bright rufous. Antennae more slender (see the key). Head less punctate inside eyes, rarely faintly rugose on frons. Frontal furrows as in fig. 124. Pronotum strongly convex to side-margins, basal foveae large. Elytra with protruding shoulders, striae finely punctate, evident also apically. Wings constantly rudimentary.

Distribution. Denmark: generally distributed and very common, except in WJ and NWJ. — Sweden: known from most districts from Sk. to Lu. Lpm., but lacking from large areas north of the river Dalälven. Common and very distributed in the south. — Norway: generally distributed north to northern Nordland (Nn). — In E. Fennoscandia north to about 65°N but scattered in north. — Europe except extreme northern and southern parts, east to Siberia and the Caucasus.

Biology. A hygrophilous species predominantly occurring in humid deciduous forests on clayish mull-soil, often near water. Also in parks and gardens and on moist, shady places in open, often cultivated land. Reproduction takes place in August-September. Larvae and a few aged adults hibernate, and the larvae pupate in the following spring. Newly emerged beetles usually occur in early summer. They are active for a short period and then enter upon an aestivation dormancy which lasts until the onset of reproduction (Thiele, 1969). Development takes two years in northern Fennoscandia.

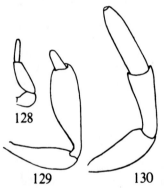

Figs 128-130. Maxillary palpi. — 128: *Perileptus areolatus* (Creutz.); 129: *Bembidion* subg. *Notaphus;* 130: *Trechus rubens* (F.).

Tribe Trechini

Small species. Separated from the Bembidiini by the well developed terminal segment of the maxillary palpi (Figs 128, 130).

A very large group with world-wide distribution, especially in mountainous districts. Predominantly subterranean, notably in caves.

Genus *Perileptus* Schaum, 1860

Perileptus Schaum, 1860, Naturg. Ins. Deutschl. Col. 1 (1): 663.
Type-species: *Carabus areolatus* Creutzer, 1799.
Blemus Laporte, 1840; *nec* Dejean, 1821.

Small, flat and narrow species. Superficially intermediate between *Bembidion* and *Trechus* through the moderately reduced terminal segment of the maxillary palp (Fig. 128). Also, the sutural stria is not "recurrent" at apex of elytra, but the frontal furrows are strongly divergent behind the eyes, as in *Trechus*. Entire upper surface with short pubescence. Penultimate segment of pro-tarsus with a long, sharp spine. Wings full. Male with two dilated segments on both pro- and meso-tarsi.

79. *Perileptus areolatus* (Creutzer, 1799)
Fig. 128; pl. 4: 17.

Carabus areolatus Creutzer, 1799, Ent. Vers., Wien: 115.

2.4-2.5 mm. Piceous, base of antennae, mouth-parts, legs and central parts of elytra pale. Head as wide as pronotum. Elytra almost parallel-sided with rectangular shoulders.

Distribution. Sweden: very rare and local, only recorded from single localities in Hall., Dlr., Hls., and two localities in Vrm. (river Klarälven). — Norway: records from several southern districts: AK, HEs, Bø, TEi, VAy, and one isolated record from STy. — Not in Denmark or Finland, but a few specimens found in Kr: Salmi. — Europe (discontinuous), N.Africa, Asia Minor, W. Siberia.

Biology. A stenotopic species, confined to gravelly-stoney river banks, at least in Norway predominently along small rivers, regularly occurring together with *Bembidion saxatile, prasinum* and *virens*. It is less hygrophilous than these species and is therefore most abundant at some distance from the water edge (Andersen, 1982). The period of reproduction is rather late in spring, females with mature eggs have been found in June.

Genus *Aepus* Samouelle, 1819

Aepus Samouelle, 1819, Ent. Comp., London: 149.
Type-species: *Aepus fulvescens* Samouelle, 1819 (= *marinus* Ström).

Very small, flat, unpigmented beetles adapted for a subterranean life in the tidal zone.

Head with rudimentary eyes (Fig. 131) and pubescent temples. Frontal furrows semi-circular, as in all Trechini. Pronotum cordiform. Elytra truncate at apex, leaving at least part of last abdominal segment free (as in the Lebiini). Elytral surface with sparse erect pubescence; striae suggested only and the sutural one recurrent at apex; two setiferous dorsal punctures. Wings completely reduced. Male with two dilated protarsal segments.

Fig. 131. *Aepus marinus* (Strøm), length 2.2-2.4 mm. (After Jeannel).

80. *Aepus marinus* (Ström, 1788)
Fig. 131; pl. 5: 1.

Carabus marinus Ström, 1788, Skr. Nye Saml. Norske Vidensk. Selsk. 2: 385.

2.2-2.4 mm. Entirely brownish yellow. Terminal segment of the maxillary palpi cylindrical.

Distribution. In Fennoscandia found on the western coast of Norway between Stavanger and Trondheim. One recent record (1982) from Sweden: Boh., Råssö, where the species was rather frequent during the summer. — Also the British Isles, Ireland, France (Britanny, Normandy).

Biology. Restricted to rocky seashores in the tidal zone, in Norway confined to shallow creeks, occurring under big stones on sand. Its habitat is usually below the high-water mark and is therefore flooded by the tide. In England the species has also been found above the high-water mark, in *Zostera*-heaps. The beetles are very shy of light and apparently do not leave their dwelling places (Andersen, 1960). Richoux (1972) found no morphological or physiological adaptations to submersion. During flooding the beetles seek out crevices into which the water do not enter. Females of *Aepus*

116

lay only one large egg at a time. *A.marinus* sometimes occurs together with the staphylinid *Micralymna marinum* Ström. It has been found numerous in September.

Genus *Trechus* Clairville, 1806

Trechus Clairville, 1806, Ent. Helv. 2: 22.
Type-species: *Trechus rubens* Clairville, 1806 (*nec* Fabricius) (= *Carabus quadristriatus* Schrank, 1781.
Blemus Dejean, 1821, Catal. Coleopt.: 16.
Type-species: *Carabus discus* Fabricius, 1792.
Epaphius Stephens, 1827, Ill. Brit. Ins. Mand. 1: 67.
Type-species: *Carabus secalis* Paykull, 1790.
Lasiotrechus Ganglbauer, 1892, Käf. Mitteleur. 1: 187.
Type-species: *Carabus discus* Fabricius, 1792.
Trechoblemus Ganglbauer, 1892, Käf. Mitteleur, 1: 187.
Type-species: *Carabus micros* Herbst, 1784.

In its widest sense (as here taken) this is a very large genus with world-wide distribution. The North European species fall into different subgenera as follows: *Epaphius* with the species *T. secalis* and *T. rivularis, Trechus* s.str. with the species *T. rubens, T. fulvus, T. quadristriatus* and *T. obtusus, Trechoblemus* with *T. micros* and *Blemus* (of which *Lasiotrechus* is a synonym) with *T. discus*. Occasionally these subgenera are considered separate genera.

Its members are small, in general habitus (Fig. 132) and movements similar to *Bembidion*. The ground colour is demelanised, yellowish to piceous, never pure black, and the appendages are pale. Metallic colours are absent, but the elytra are often more or less iridescent which is caused by a dense, transverse microsculpture. The most important distinguishing characters are: (a) the well developed terminal segment of the maxillary palp (Fig. 130); (b) the backwardly strongly diverging frontal furrows; (c) the "recurrent" sutural stria of the elytra (Fig. 133); this last feature is, however, also present in *Tachys* and subgenus *Ocys* of *Bembidion*. Third elytral interval with three dorsal punctures; the abbreviated scutellar stria obsolete or lacking. Wings varying. Male with two dilated pro-tarsal segments. The internal sac of penis with more or less well developed sclerites. Parameres almost symmetrical with long apical setae.

Our species usually occur in more or less shady places and are moderately hygrophilous; none is riparian. Some species (e.g. *T. rivularis* and *quadristriatus*) are known to be carnivorous, preying on Collembola, insects eggs, mites, etc.

Key to species of *Trechus*

1 Elytra with short but dense pubescence over entire surface 2
–   Elytra glabrous (except for dorsal punctures) 3
2 (1) Pronotum also with fine, depressed pubescence. Eyes

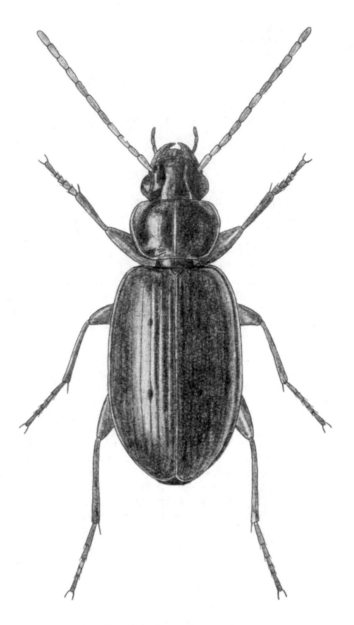

Fig. 132. *Trechus quadristriatus* (Schrk.), length 3.5-4 mm. (After Victor Hansen).

118

small, their diameter not exceeding distance to antennal insertion .... 87. *micros* (Herbst)
- Pronotum glabrous. Diameter of eyes more than twice as long as distance to antennal insertion .... 88. *discus* (Fabricius)
3 (1) Base of pronotum (Fig. 137) sinuate laterally, hind-angles very obtuse, almost obsolete .... 81. *secalis* (Paykull)
- Base of pronotum straight or only slightly sinuate laterally, hind-angles angulate or denticulate 4
4 (3) Elytra with at least 6 well developed and punctate striae 5
- Elytra with 3 or 4 evident, almost imperceptibly punctate striae 6
5 (4) Eyes small and flat, their diameter not exceeding distance to antennal insertion. Elytra not iridescent .... 84. *fulvus* Dejean
- Eyes protruding, diameter much longer than distance to antennal insertion. Elytra strongly iridescent .... 83. *rubens* (Fabricius)
6 (4) More than 4 mm. Base of pronotum straight (Fig. 138), hind-angles sharp, denticulate. First and second elytral striae parallel to apex........................... 82. *rivularis* (Gyllenhal)
- Usually less than 4 mm. Base of pronotum oblique laterally (Fig. 136), hind-angles somewhat blunt. First and second elytral striae diverging before apex 7
7 (6) Wings constantly full. Head more or less markedly darker than elytra. Hind-angles of pronotum more pronounced (Fig. 136) .... 85. *quadristriatus* (Schrank)
- Wings highly reduced (with very few individual exceptions). Entire upper surface uniformly dark. Hind-angles of pronotum more or less obsolete .... 86. *obtusus* Erichson

81. **Trechus secalis** (Paykull, 1790)
   Fig. 137.

Carabus secalis Paykull, 1790, Mon. Car. Suec.: 94.

3.5-4 mm. A small species, at once recognized on the structure of the pronotum (Fig. 137). Uniformly testaceous or rufous, or with elytra infuscated apically. Pronotum with very obtuse hind-angles. Elytra with rounded shoulders and inner striae strongly punctate. Wings constantly reduced into small scales.

Distribution. Denmark: widely distributed and rather common except in WJ. — Sweden: in almost every district from Sk. to Ly. Lpm. and very distributed and common in the south. — Norway: north to 68°N in TRi. — Finland: north to 65°N in Ob. — Also in C. Europe.

Biology. In moist, rather shaded sites, under dead leaves and other debris. Mostly on clay soil rich in humus, less often on sandy or peaty soils, in woodland (mainly water-meadow forests) as well as in open country (notably in W. and N. Fennoscandia) in rich meadows and on arable land. The species is most frequent in June-September. Newly emerged beetles occur in early summer, and reproduction takes place in the autumn.

82. *Trechus rivularis* (Gyllenhal, 1810)
    Fig. 138; pl. 5: 4.

Bembidium rivularis Gyllenhal, 1810, Ins. Suec. 2: 33.

4.4-4.8 mm. Broad and convex, superficially similar to a *Bembidion* of the subg. *Philochthus*, but with a characteristic pronotum. Our darkest species, rufo-piceous, elytra darker and iridescent. First antennal segment and legs palest. Base of pronotum (Fig. 138) straight, hind-angles dentiform. Elytra with 1st and 2nd striae parallel to apex, posterior dorsal puncture more removed from apex than in any other species. Wings dimorphic, either full or reduced into a tiny scale.

Distribution. Denmark: very rare; not in Jutland and altogether 12 records from the islands. — Sweden: Sk., Ög. to Lu. Lpm.; rare and very local, absent from large

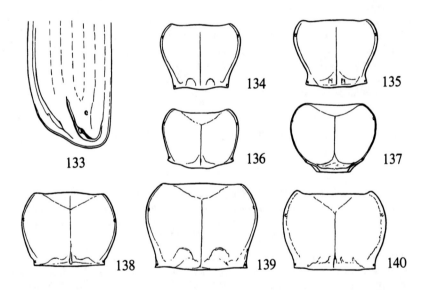

Fig. 133. Elytral apex of *Trechus,* with recurrent stria.
Figs 134-140. Pronotum of *Trechus.* — 134: *micros* (Hbst.); 135: *discus* (F.); 136: *quadristriatus* (Schrk.); 137: *secalis* (Payk.); 138: *rivularis* (Gyll.); 139: *rubens* (F.); 140: *fulvus* Dej.

areas, especially north of the river Dalälven. — Norway: only AK and NTi. — Finland: rather rare, in the southern and central districts; also Vib and Kr in the USSR. — N. Europe and northern C.Europe.

Biology. In peaty woods with growth of birch, alder or willow and with a vegetation dominated by *Sphagnum, Carex,* ferns, etc. The species occurs on dark, moist sites among wet leaves, usually between pillows of *Sphagnum;* often together with *Agonum livens* and *Patrobus assimilis.* It is most frequent in June-July. Newly emerged beetles occur in June. Reproduction takes place in summer, and the larvae hibernate.

83. *Trechus rubens* (Fabricius, 1792)
Figs 130, 139.

Carabus rubens Fabricius, 1792, Ent. Syst. 1: 140.

5-6.5 mm. Entirely reddish brown, head somewhat darker, elytra often paler and strongly iridescent from very dense transverse microsculpture. Eyes large and protruding (see the key). Pronotum (Fig. 139) with acute hind-angles. Elytra with 8 punctate striae, the three outer of which are obsolete. Wings full and often used.

Distribution. Denmark: very rare; scattered distributed in Jutland; on the islands only 19th Century records from SZ, NEZ and B. — In Fennoscandia generally distributed except in the extreme north. — C. and N.Europe. Introduced in North America.

Biology. A hygrophilous, more or less subterranean species. Found in open country, in shaded places with rather dense vegetation as well as in moist forest; often on the banks of rivers and brooks and near human habitations. The species occurs on clayey, mull-rich and on peaty soil, usually among dead leaves and other debris or under big stones. It flies often at night and has been observed to occur numerous in sea drift. *T. rubens* in most predominant in May-September and is considered a spring breeder (Lindroth, 1945). Findings of newly emerged beetles in June suggest that autumn breeding may also occur.

84. *Trechus fulvus* Dejean, 1831
Fig. 140.

Trechus fulvus Dejean, 1831, Spec. Gén. Col. 5: 10.
Trechus lapidosus Dawson, 1849, Ann. Mag. Nat. Hist. (2) 3: 214.
Trechus Rathkei Helliesen, 1892, Stavanger Mus. Aarsh. 1890-92: 31.

4.8-5.7 mm. In size and general outline similar to *rubens,* but flatter and paler and without any trace of iridescense. Entirely testaceous. Eyes small and flat (see the key). Sides of pronotum (Fig. 140) deplanate to front-angles. Elytra without microsculpture, parallel-sided, apex rather abruptly truncate, striae strong. Wings atrophied.

Distribution. Not in Denmark. — In Fennoscandia only known in Norway: Ry and STy. — Shetland, Scotland, England, S.Wales, Ireland, Helgoland, France, Iberian Peninsula.

Biology. On rocky sea-shores under stones near the high-water mark. A separate subspecies inhabits caves on the Iberian Peninsula.

85. *Trechus quadristriatus* (Schrank, 1781)
   Figs 132, 136, 142, 144.

Carabus quadristriatus Schrank, 1781, Enum. Ins. Austr.: 218.

3.5-4 mm. A small, pale species. Testaceous to brown, head and abdomen darkest, elytra slightly iridescent, usually with paler shoulders. Anterior supra-orbital puncture close to eye. Elytra with 3-4 evident and 2 suggested striae, 2nd stria diverging from 1st before apex. Wings constantly full and often used in specimens from our area, but a brachypterous form is known from S.Europe. Penis (Figs 142, 144) straight with short apex; internal sac without defined sclerites.

Distribution. In Denmark generally distributed and common. — Sweden: very distributed and common in the south, north to Jmt., Âng. and Nb. — Norway: only in the southern parts, north to Os. — Finland: north to ca 64°N in Om; also Vib and Kr in the USSR. — Europe and W.Asia.

Biology. Predominantly on open, rather dry ground with short vegetation of grasses etc., usually on sandy or gravelly, often clay-mixed soil. It is a common inhabitant of sea-dunes, occurring together with e.g. *Dromius linearis,* and is also frequently en-

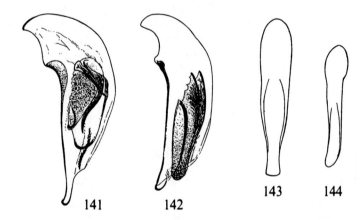

Figs 141-144. Penis of *Trechus,* lateral and dorsal views. — 141 & 143: *obtusus* Er.; 142 & 144: *quadristriatus* (Schrk.).

countered on arable land, mainly in root crops. In C.Europe also in forest and in moist shore habitats. The species prefers somewhat shaded ground, in cabbage fields thus occurring under the plants (Mitchell, 1963b). The adults prey upon insect eggs and may play an important role in the natural control of certain insect pests, among them the cabbage root fly. *T. quadristriatus* is found almost throughout the year, but mainly in July-August. It is primarily an autumn breeder having winter larvae, but some adults may breed in spring (Mitchell, 1963a).

86. *Trechus obtusus* Erichson, 1837
Figs 141, 143.

Trechus obtusus Erichson, 1837, Käf. Mark Brandenburg 1 (1): 122.

3.6-4.1 mm. Very similar to *quadristriatus*. Somewhat broader with sides of elytra a little more rounded. Colour of upper surface more uniformly greyish brown. Anterior supra-orbital puncture more distantly removed from eye, which is somewhat smaller and flatter. Pronotum with hind-angles less evident and margins more rounded. The hind-wings are in our individuals almost constantly reduced into a scale much shorter than elytra, but fully winged individuals have been observed. Penis (Figs 141, 143) with very broad base and prolonged apex; internal sac with spine-like sclerites.

Distribution. Denmark: widely distributed and common. — Sweden: rare, only known from a few localities in two widely separated areas. In the south recorded to high elevated areas in Hrj., Jmt., Ås. Lpm. and Ly. Lpm. — Norway: scattered records from several districts north to c. 68°N. — Not in E. Fennoscandia. — Also C. and S.Europe.

Biology. Less xerophilous than the preceding species, occurring in moderately humid and usually shaded sites, for instance in hedges and thin deciduous forest. In England, Pollard (1968) found *T. obtusus* in hedges, while *quadristriatus* was almost confined to the surrounding fields. In northern Fennoscandia the species mainly occurs in open country, for instance in the subalpine and lower alpine regions of the mountains, but also in coastal habitats. The species seems to reproduce in both spring and autumn.

87. *Trechus micros* (Herbst, 1784)
Fig. 134; pl. 5: 3.

Carabus micros Herbst, 1784, Arch. Insectengesch. 5: 142.

4-4.5 mm. Easily recognized by the presence of pubescence of both pronotum and elytra. Very narrow. Dull testaceous; centre of head, and usually also an oblong diffuse spot on each elytron, darker. Eyes small (see the key). Pronotum: fig. 134. Recurrent elytral stria at apex joining with 3rd stria, not with 5th stria as in all preceding species. Wings full.

Distribution. Denmark: scattered distributed in E. and C.Jutland and on the islands (not B); only single records from WJ and NWJ. — Sweden: rare and local but will probably prove to be more distributed. Recorded from Sk. to Upl., Vstm., Vrm., Med., Ång., Vb.; not Öl. and Gtl. — Norway: some south-eastern districts and STi and STy. — In E. Fennoscandia in the extreme south of Finland and in Vib and Kr of the USSR. — Entire Europe except the south. N.Asia.

Biology. Always found near fresh water, often occurring on the banks of rivers and lakes and in wet meadows, usually on clay-mixed peaty soil. The species is pronouncedly subterranean and is regularly found in the burrows of rodents and moles; often together with *T. discus*. After inundation, for instance after show-melting in early spring, *T. micros* is forced out of its dwelling-places and may be taken in abundance (Baranowski & Sörensen, 1978). The species is most numerous in spring and early summer when breeding probably takes place. The adults hibernate.

88. *Trechus discus* (Fabricius, 1792)
    Fig. 135; pl. 5: 2.

Carabus discus Fabricius, Ent. Syst. 1: 164.

4.4-5.5 mm. Only elytra pubescent. Eyes of normal size (see the key). Broader than *micros,* ground colour more rufous, dark spot of elytra transverse. Pronotum (Fig. 135) with hind-angles more protruding. Elytra more shiny, their pubescence less depressed. Wings full and functional.

Distribution. Denmark: scattered and rare in eastern Jutland and on the islands (also B); single records from WJ, NWJ and NEJ. — Sweden: in most districts from Sk. to Hls.; rare, absent from large areas; generally distributed only along the west coast. — Norway: Ø, AK, VE, STi and NTi. — In E. Fennoscandia in the southern part of Finland (N to 62°N) and in Vib and Kr of the USSR. — C. and N.Europe, N.Asia. Introduced in North America.

Biology. The habitat preference of this species is identical to that of *T. micros*. It prefers clayey soil with rather dense vegetation of grasses, sedges, etc., and occurs usually near water, for instance in eutrophic fens at the margins of lakes and rivers. Like the preceding species it is decidedly subterranean and has been found in the burrows of rodents and moles. The species can be taken in great number after inundation of its habitat, sometimes in company with *T. micros*. It is an autumn breeder having winter larvae; adult beetles are most numerous in August.

Tribe Bembidiini

The rudimentary terminal segment of the maxillary palpi is the diagnostic character of this tribe. Represented by four genera in our area.

124

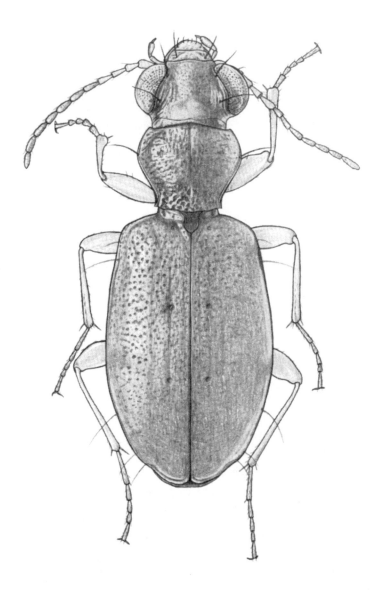

Fig. 145. *Asaphidion flavipes* (L.), length 3.9-4.7 mm.

Genus *Asaphidion* des Gozis, 1886

Asaphidion des Gozis, 1886, Res. Esp. Typ.: 6.
Type-species: *Cicindela flavipes* Linnaeus, 1761.
Tachypus Dejean, 1821, Cat. Coll. Col. B. Dejean: 16; *nec* Weber, 1801.

At once separated from other genera of the tribe by the irregularly punctate and pubescent elytra; the striae indicated only near the suture as indistinct furrows. Due to the enormous eyes, the general habitus reminds of *Elaphrus* and also the movements are similar. Pronotum cordiform with sharp hind-angles. Elytra with 2 dorsal punctures near the suture. Wings full. Male with first pro-tarsal segment strongly, second faintly, dilated.

The members of the genus are hygrophilous, usually occurring near water; they are diurnal, visually hunting beetles.

Key to species of *Asaphidion*

1 Length 5.0-6.0 mm. Head not wider than pronotum, which
lacks a latero-basal keel .... 89. *pallipes* (Duftschmid)
– Length 3.9-4.7 mm. Head with very convex eyes and therefore wider than pronotum, which has a fine oblique keel at
hind-angles 2
2 (1) Appendages rufo-testaceous, but the antennae are slightly
darker apically and the knees greenish metallic infuscated.
Upper surface usually with a bronzy lustre. Male and female
genitalia (Figs 146a, 147a) .... 90. *flavipes* (Linnaeus)
– Antennae and legs entirely pale. Upper surface normally
with a red coppery lustre. Male and female genitalia
(Figs 146b, 147b) .... 91. *curtum* Heyden

89. *Asaphidion pallipes* (Duftschmid, 1812)
Pl. 5: 5.

Elaphrus pallipes Duftschmid, 1812, Fauna Austriae 2: 197.

5.0-6.0 mm. Black, upper surface coppery with golden or bluish spots and stripes, especially laterally on elytra. Pubescence forming patches of yellowish or silvery grey. Appendages pale but first and outer antennal segments as well as femora and tarsi infuscated, with greenish hue. Punctuation of elytra extremely fine and dense.

Distribution. Denmark: rare, but rather distributed, and seemingly absent from large areas of the western part of Jutland; sparse occurrence in F and NWZ. — Sweden: scattered occurrence all over the country but usually rare. Very local in the south, more generally distributed in the eastern part of Norrland. Not recorded from

Bl., Hall., G. Sand., Boh., Dlsl., Sdm., Gstr., and Hrj. — Norway: scattered through the south-eastern, central and northern districts, locally common. No records from VE, AA, VA, R, HO and SF. — Finland: distributed over most of Finland, and locally common. No records from Al and some northern districts. — USSR: many records from the Karelian Isthmus and the Svir area; otherwise few specimens. — Europe, the Caucasus and western Siberia.

Biology. On slightly moist fine sand or clay; in open country among sparse vegetation, e.g. patches of moss, as well as in habitats somewhat shaded by scattered bushes. In C. and N. Fennoscandia predominantly on river banks, in southern Scandinavia

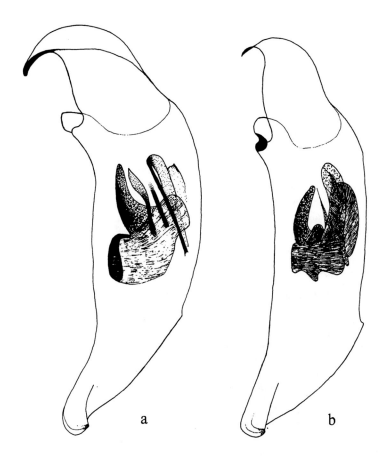

Fig. 146. Penis of a: *Asaphidion flavipes* (L.) and b: *A. curtum* Heyd. Viggo Mahler del.

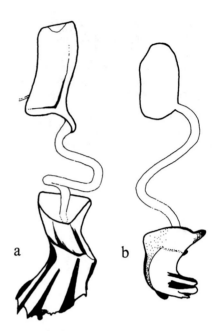

Fig. 147. Spermatheca and Annulus receptaculi of a: *Asaphidion flavipes* (L.) and b: *A. curtum* Heyd. Viggo Mahler del.

a

b

mainly on sea shores and slopes and in gravel or clay pits; more sporadic in forest edges and on formerly cultivated soil. Populations of *A. pallipes* may consist of spring and autumn breeding groups of animals (Andersen 1970).

90. *Asaphidion flavipes* (Linnaeus, 1761)
Figs 146a, 147a.

Cicindela flavipes Linnaeus, 1761, Fauna Suec. ed. 2: 211.

3.9-4.7 mm. Smaller, flatter and with broader elytra than *pallipes*. Metallic lustre of upper surface less pronounced, bronzy, less spotty. All appendages pale, except that the antennae are slightly infuscated apically and the knees are greenish metallic darkened. Eyes more produced than in *pallipes* and head therefore wider than pronotum which has a fine, oblique keel inside hind-angles. Punctuation of upper surface more coarse and sparse than in *pallipes*. Penis, fig. 146a. Female genitalia, fig. 147a.

Distribution. Denmark: rather distributed in SJ, EJ and the eastern provinces, absent from great parts of northern, western and central Jutland. — Sweden: generally distributed and sometimes common in the south to 61°. North of 61°N very rare and local, only a few records from N. Dalarna, Hls., and Äng. — Norway: rare, only recorded from Ø, AK, HE, B and VE. — Finland: distributed over southern and central Finland (not Al), and locally common. — USSR: locally common in the southern part of the area covered. — Europe, the Caucasus and western Siberia.

Biology. Seemingly preferring open ground with sparse vegetation, usually moist clayey or clay-mixed sandy soil near water, for instance occurring on river banks, sea slopes, in clay pits, also on cultivated soil. Breeding takes place in spring. Eggs are laid singly in moist soil (Bauer 1971).

91. *Asaphidion curtum* Heyden, 1870
Figs 146a, 147b.

Asaphidion curtum Heyden, 1870, Reise Spanien, Berlin, p. 65.

3.9-4.3 mm. Closely related to, and earlier often regarded as a variant of, *flavipes* but the male and female genitalia are quite different (Focarile 1964, Lohse 1983). Appendages entirely pale. Antennae relatively shorter than in *flavipes*. Sides of pronotum mostly sharply angular, in *flavipes* more rounded. Elytra usually with a denser punctuation and a more pronounced microsculpture, and often with deeper longitudinal furrows. The metallic colour of the upper surface is usually red coppery. Hind-wings longer than, in *flavipes* almost as long as, the elytra. Penis, fig. 146b. Female genitalia, fig 147b.

Distribution. Denmark: Rare. Very scattered occurrence in Zealand and Funen. Two records from LFM, one from EJ. No records from B. — The species can be reliable recorded from West Germany (Holstein, Hamburg, Niedersachsen), France, Spain, Italy, Algeria, and Tunisia.

Biology. In somewhat shaded sites, mainly forest habitats. In Denmark it has been found numerously on moist clayey mull-soil in deciduous forest, on open spots among vegetation of *Carex*, etc.

Note. The possible occurrence of this overlooked species in Sweden could not be verified (R. Baranowski *in litt.*).

Genus *Bembidion* Latreille, 1802

Bembidion Latreille, 1802, Hist. Nat. Crust. Ins. 3: 82.
  Type-species: *Carabus quadrimaculatus* Linnaeus, 1761.
Bembidium auctt.
Ocydromus Clairville, 1806, Ent. Helv. 2: 20.
  Type-species: *Carabus modestus* Fabricius, 1801.
Odontium LeConte, 1848, Ann. Lyc. Nat. Hist. New York 4: 451.
  Type-species: *Bembidion coxendix* Say, 1823.
For other generic names, see the subgenera.

This is the largest Carabid genus, notably in temperate and arctic regions; in the tropics its role is largely taken over by the *Tachys* group. There have also been many attempts to divide *Bembidion* into smaller genera, so e.g. by Jeannel (1941), who recognized 17 genera, with sometimes rather insignificant differences. A more balanced arrangement was recently proposed by Perrault (1981). As divided by him, the North European *Bembidion* species would fall into 7 genera: *Odontium, Metallina, Phyla, Ocys, Bembidion, Cillenus* and *Ocydromus. Odontium* would include the species numbered 92-96, *Metallina* 97-100, *Phyla* 101, *Ocys* 102-103, *Bembidion* 108-136, *Cillenus* 137 and *Ocydromus* those numbered 104-107 and 138-164. So far, however, no arrangement has won general acceptance.

All species are small (not exceeding 7.5 mm) with slender appendages and very fast movements. The upper surface is usually more or less metallic but often with pale markings on the elytra.

The most important diagnostic character is the rudimentary form of the terminal segment of the maxillary palpi (Fig. 129), as in the three other members of the tribe *(Asaphidion, Tachys, Tachyta)* which are separated from *Bembidion* as described under each of these genera. Among other Fennoscandian ground-beetles, only the genus *Perileptus* has a similar, but less pronounced reduction of the palpi (Fig. 128). The other members of the tribe Trechini are also superficially similar, but their frontal furrows, among other things, are semicircularly prolonged backwards behind the eye.

Most species are fully winged but several are dimorphic or constantly brachypterous. The male has 2 strongly dilated pro-tarsal segments. The penis, with its often extremely complex armature of the internal sac, is often completely decisive for species determination. The organ should be treated according to the device given in the introduction. The parameres are asymmetric, the left one broad and triangular.

The separation of species is often difficult, not only due to their excessive number, but also because many of them are extremely similar or vary individually within the species.

External characters of particular taxonomic importance are: *Frontal furrows* (Figs 148-156), a pair of more or less well-defined sulci along the inside of each eye. They may be doubled. *Supra-orbital punctures,* 2 on each side, inside the eye. *Dorsal punctures* (setigerous) usually 2nd on 3rd interval or attached to 3rd stria. *Preapical puncture* of elytra (setigerous), near the apex of elytra, closer to side-margin than to suture (Fig. 191). *Preapical spot* of elytra, laterally near apex. It is sometimes not visible unless the elytron is lifted.

A study of the microsculpture, notably of the elytra, is of outstanding value in most subgenera. It is often stronger in the ♀, and specimens of the same sex must always be compared. Width and length of microsculpture "meshes" are determined according to the long axis of the insect.

In order to fix the taxonomic status of each species, they have been referred to the generally recognized subgenera below. This division essentially follows Netolitzky (1943).

Most *Bembidion* are strongly hygrophilous and live close to water, where they run about with great agility. Some are confined to running waters, others to shores of lakes, ponds, or the sea. They are often dependent upon a special kind of soil.

Most species are primarily scavengers, but many also take living prey such as small arthropods, insect eggs, etc. They are predominantly spring breeders hibernating as imagines, but some species (e.g. *B. hastii*) may also hibernate as larvae or pupae; a few (e.g. *B. lunatum*) are true larval hibernators. Spring and early summer are generally the best period for collecting. The biology of several northern, riparian species of *Bembidion* has been described by Andersen (1968, 1970, 1978).

Key to species of *Bembidion*

1	3rd elytral interval much broader than 2nd and 4th at middle, with two well defined opaque fields ("silver-spots") (Fig. 188)	2
–	3rd elytral interval not different from adjacent ones	5
2(1)	4th elytral stria suddenly bent in front of "silver-spots"; outer intervals (at least 7th) with alternating dull and shiny fields	95. *litorale* (Olivier)
–	4th elytral stria not bent; outer intervals uniformly dull	3
3(2)	Pronotum (Fig. 160) with rather obtuse hind-angles; only posterior marginal seta (at hind-angle) present	96. *argenteolum* Ahrens
	Hind-angles of pronotum right or acute; 2 marginal setae (the anterior at middle) present	4
4(3)	1st antennal segment (also above) and base of the 3-4 following, tibiae (except knees) and tarsi pale rufous (though sometimes with metallic hue). Pronotum (Fig. 158) with sides more rounded and hind-angles less protruding. Penis, Fig. 194	93. *velox* (Linnaeus)
–	Only 1st antennal segment (and usually only underneath) pale; legs entirely dark or tibiae and base of femora dark rufous. Penis, Fig. 195	94. *lapponicum* Zetterstedt
5(1)	Elytra at middle with 9 equal complete evidently punctate striae (abbreviated scutellar stria not counted). 5.5-6.5 mm	92. *striatum* (Fabricius)
–	If 9 complete striae present, the 8th is impunctate, more impressed (especially apically) and almost fused with 9th stria	6
6(5)	Sutural stria recurrent at apex (as in *Trechus*, Fig. 133). Dorsal punctures (1 or 2) situated behind middle	7
–	Sutural stria not recurrent. At least foremost dorsal puncture before middle	8
7(6)	Base of pronotum oblique laterally (Fig. 165). Elytra with 2 dorsal punctues	103. *quinquestriatum* Gyllenhal

val free, not touching any of the adjacent striae 19
- Dorsal punctures touching 3th stria or situated within it 28
19(18) Elytra with striae disappearing before apex, entirely dark
or more or less pale in posterior half. Small (2.3-3.2 mm) 20
- Elytra with striae complete to apex, usually pale also in anterior half . . . 23
20(19) Frontal furrows virtually parallel (Fig. 151), obsolete on clypeus 21
- Frontal furrows converging, at least anteriorly (Figs 152,
153) and evident also on clypeus 22
21(20) Pronotum pronouncedly cordiform (Fig. 176), clearly
wider than head . . . . 114. *minimum* (Fabricius)
- Pronotum less widened forwards (Fig. 177), narrower. Ely-
tral striae with stronger punctures . . . . 115. *normannum* Dejean
22(20) Pronotum (Fig. 181) with base oblique inside the obtuse
hind-angles. Frontal furrows almost parallel between the
eyes (Fig. 153) . . . . 117. *tenellum* Erichson
- Pronotum with almost straight base, hind-angles about
right (Fig. 178). Frontal furrows generally converging . . . .
. . . . 116. *azurescens* (Dalla Torre)
23(19) Entire upper surface without microsculpture very shiny.
Ground colour of elytra pale . . . . 113. *ephippium* (Marsham)
- Upper surface microsculptured, at least forebody dull.
Ground colour of elytra dark 24
24(23) Elytra iridescent due to the microsculpture which consists
of extremely fine and dense transverse lines. Anterior su-

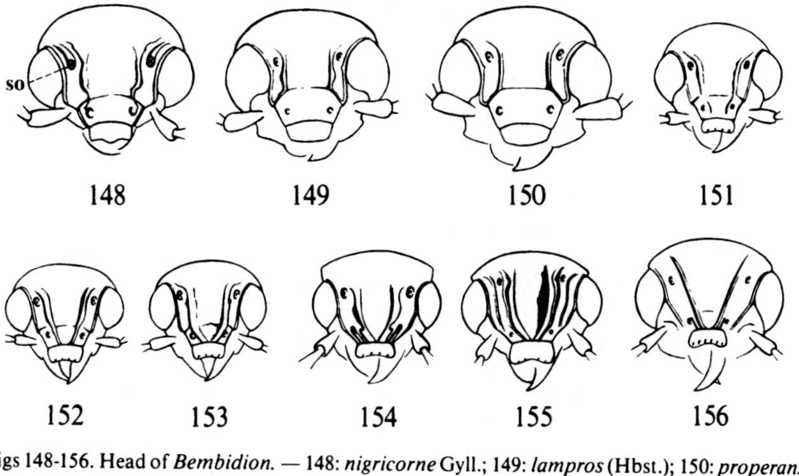

Figs 148-156. Head of *Bembidion*. — 148: *nigricorne* Gyll.; 149: *lampros* (Hbst.); 150: *properans* Stph.; 151: *minimum* (F.); 152: *azurescens* (D. Torre); 153: *tenellum* Er.; 154: *schuppelii* Dej.; 155: *assimile* Gyll.; 156: *doris* (Pz.).

133

pra-orbital puncture surrounded by an elevated shiny field 25

– Elytra micro-reticulate, not iridescent. Frons without shiny field 26

25(24) Elytra with evident shoulder-tooth; their pale pattern diffuse. Pronotum with broader base. Penis, Fig. 200 . 109. *tinctum* Zetterstedt

– Shoulder-tooth suggested only; pale spots well defined. Pronotum, Fig. 166. Penis, Fig. 199 .... 108. *dentellum* (Thunberg)

26(24) Antennae with 3 or 4 pale segments (though sometimes with metallic hue on dorsum). Meshes of elytral microsculpture somewhat irregular............. 111. *semipunctatum* (Donovan)

– At most 1st antennal segment pale, 2nd-4th at least dark dorsally. Micro-meshes of elytra regular, arranged as bricks 27

27(26) Larger (4.1-5.1 mm). Pronotum broader with more rounded sides. Elytra almost parallel-sided at middle. Apex of elytra pale, legs brownish .... 110. *varium* (Olivier)

– Smaller (3.0-4.4 mm). Elytra with sides more rounded, somewhat widening posteriorly. Normally with dark elytral apex and legs almost black .... 112. *obliquum* Sturm

28(18) Hind-angles or pronotum very sharp removed from the level of the base and separated from it by an incision (Figs 179, 180). Elytra with 1 or 2 pale spots 29

– Hind-angles not or very little removed from base 31

29(28) Antennae with 4 pale basal segments. Legs entirely pale or with femora slightly infuscated. Pronotum, Fig. 179.................... .... 128. *quadrimaculatum* (Linnaeus)

– Antennae and femora black (or dark piceous) 30

30(29) Elytra with a single orange spot behind shoulder. 2.6-3.0 mm .... 129. *humerale* Sturm

– Elytra also with pale subapical spot. 3.5-4.0 mm .... 130. *quadripustulatum* Audinet-Serville

31(28) Frontal furrows sharply prolonged upon clypeus to base of labrum, either doubled (entirely or anteriorly, Fig. 156) or strongly convergent (Fig. 155). Not over 4 mm...................... 32

– Frontal furrows usually shallow, never prolonged upon clypeus, and more and less parallel (Fig. 154). Usually larger 40

32(31) Frontal furrows not doubled, straight, strongly converging (Fig. 156).... 33

– Frontal furrows doubled, at least anteriorly parallel on frons, converging on clypeus (Fig. 155) 35

33(32) Pronotum (fig. 184) on each side between fovea and median line with a single small impression. Elytra dark, usually black, and with a pale preapical spot .... 120. *doris* (Panzer)

– Pronotum (Fig. 182) on each side with 2 small impressions inside basal fovea. Elytra with pale spots also in anterior half.......... 34

34(33) 2.9-3.9 mm. Pronotum (Fig. 182) at base narrower than head behind eyes. Pale spots of elytra confluent .. 118. *articulatum* (Panzer)

–	2.5-2.8 mm. Pronotum at base wider than head behind eyes.

– 2.5-2.8 mm. Pronotum at base wider than head behind eyes.
   All pale spots of elytra distinct . . . . 119. *octomaculatum* (Goeze)
35(32) Frontal furrows doubled for their entire length (Fig. 155) 36
– Frontal furrows doubled anteriorly only 39

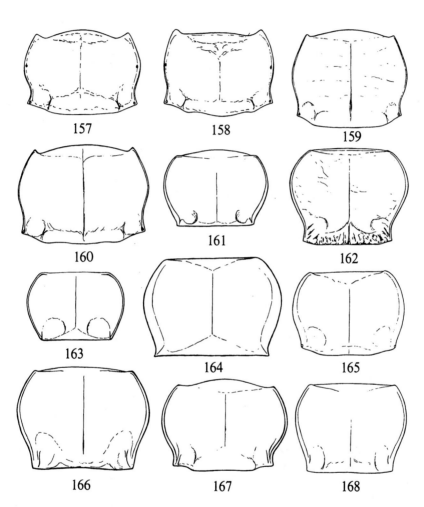

Figs 157-168. Pronotum of *Bembidion*. — 157: *lapponicum* Zett.; 158: *velox* (L.); 159: *litorale* (Oliv.); 160: *argenteolum* (Ahr.); 161:*nigricorne* Gyll.; 162: *mckinleyi* Fall; 163: *obtusum* A.-Serv.; 164:*harpaloides* A.-Serv.; 165: *quinquestriatum* Gyll.; 166: *dentellum* (Thbg.); 167: *varium* (Oliv.); 168: *monticola* Sturm.

36(35) Pronotum with microsculpture obsolete on disc and therefore shiny. Apex of elytra dark .... 37.
   –   Pronotum densely microsculptured over entire surface and
therefore dull. Apex of elytra pale 38
37(36) Preapical spot of elytra usually clearly defined. Microsculpture of elytra strong, in the female forming evident,
transverse meshes .... 127. *transparens* (Gebler)
   –   Preapical spot of elytra more diffuse. Microsculpture of
elytra irregularly transverse without evident meshes at
apex .... 126. *clarkii* (Dawson)
38(36) 3.5-4.0 mm. Elytra with distinct pale spots also in basal
half. Striae shallower. Wings full .... 124. *fumigatum* (Duftschmid)
   –   2.8-3.5 mm. Elytra with basal half immaculate or with
somewhat indistinct spots. Wings often reduced ... 125. *assimile* Gyllenhal
39(35) Upper surface unmetallic. Legs entirely pale. Elytra shiny
due to lack of microsculpture, strial punctures very coarse
anteriorly. 2.5-3.0 mm .... 123. *gilvipes* Sturm
   –   Upper surface metallic or iridescent, elytra dull from transverse microsculpture. Strial punctuation moderate 41
40(31) 4.0-4.2 mm. Elytral apex broadly but somewhat diffusely
rufescent; their microsculpture consisting of extremely fine
and dense, not confluent lines .... 122. *chaudoirii* Chaudoir
   –   2.8-3.2 mm. Elytra dark with metallic lustre; their microsculpture lines coarsely confluent .... 121. *schuppelii* Dejean
41(39) Elytral striae (except 1st) disappearing behind middle; elytra with 2 strongly defined pale spots, shiny because of
lacking microsculpture. Pronotum (Fig. 183) hardly broader than head .... 138. *genei* Küster
   –   Elytra with at least inner striae evident in apical half; microsculpture present. Head narrower than pronotum 42
42(41) All elytral striae evident, or almost so, to apex; 7th not weaker than 6th . 43
   –   Elytral striae usually obsolete or evanescent near apex; 7th
rudimentary or absent 50
43(42) 7th elytral stria stong, reaching apex. Latero-basal carina
of pronotum very thin or quite rudimentary .. 146. *hirmocaelum* Chaudoir
   –   7th elytral stria evident, but not fully reaching apex. Latero-
basal carina of pronotum sharp 44
44(43) Marginal bead of elytra at shoulder connected with the
abbreviated basal border, which has a forwardly directed
convexity (Fig. 189) 45
   –   Marginal bead of elytra continuing in even curvature
around shoulder (Fig. 190) 46
45(44) Elytral microsculpture is at least near the apex, and especially in the ♀, reticulate (Fig. 210). Pronotum more con-

stricted basally. Penis, Fig. 212 . . . . 139. *fellmanni* Mannerheim
  –   Elytral microsculpture (Fig. 211) not clearly reticulate. Pro-
      notum broader with sides less constricted 47
46(44) Pronotum less convex; with microsculpture reticulate (al-
      most as in *fellmanni*) and base slightly sinuate laterally.
      Penis, Fig. 214 . . . . 141. *crenulatum* F. Sahlberg
  –   Disc of pronotum with less evidently reticulate microsculp-
      ture, base not sinuate laterally. Penis, Fig. 213 . . 140. *difficile* (Motschulsky)
47(45) Abdominal sternites with a fringe of bristles along hind-
      margin (Fig. 192). Elytral striae evidently punctate 48
  –   Abdominal sternites only with the usual single pair of setae
      (Fig. 193)[1]. Elytral striae almost impunctate 49

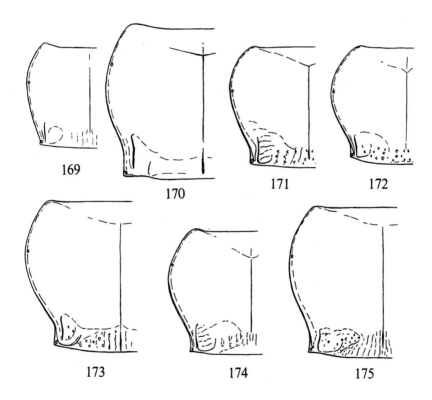

Figs 169-175. Pronotum of *Bembidion*. — 169: *tibiale* (Dft.); 170: *hastii* Sahlbg.; 171: *grapii* Gyll.;
172: *dauricum* (Mtsch.); 173: *tetracolum* Say; 174: *obscurellum* Mtsch.; 175: *petrosum* Gebl.

48(47) Base of femora more or less extensively rufous. Upper surface almost constantly bluish green. Pronotum (Fig. 170)
narrower with sides more sinuate basally . . . . 145. *hastii* Sahlberg
  – Legs black. Upper surface with green or brassy lustre . 144. *virens* Gyllenhal
49(47) 1st antennal segment (at least underneath), usually also
base of femora, rufous; elytra often rufinistic. Elytral striae
virtually impunctate . . . . 142. *prasinum* (Duftschmid)
  – Appendages black. Elytral striae with small but perceptible
punctures. Rufinistic specimens not known . 143. *hyperboraeorum* Munster
50(42) Head with a group of small but sharp punctures inside and behind eyes . 51
  – Head without extra punctures 52
51(50) Pronotum smooth and shiny on disc. Elytra each with 2 di-
stinct pale rufous spots . . . . 163. *saxatile* Gyllenhal
  – Pronotum micro-reticulate on disc. Elytra dark or diffusely
rufinistic . . . . 164. *decorum* (Zenker)
52(50) Elytra uniformly dark, usually with metallic hue, some-
times diffusely rufinistic, but never with defined pale spots 53
  – Elytra with defined pale spots, as a rule one behind shoul-
der and another before apex (sometimes only the latter present), or pale with a pronounced dark transverse band about middle 60
53(52) 2nd elytral stria as stong as 1st at apex. Base of pronotum
impunctate . . . . 147. *tibiale* (Duftschmid)
  – 2nd stria weaker than 1st towards apex, often irregular or obsolete 54
54(53) Pronotum with fine micro-meshes over its entire surface,
though sometimes less evident on disc 55
  – Pronotum very shiny, microscopically smooth on centre 56
55(54) Microsculpture of elytra consisting of evident, transverse
meshes. Basal antennal segments piceous . . . . 148. *mckinleyi* Fall
  – Elytral microsculpture densely transverse without forming
evident meshes. 1st antennal segment clear rufous . . . 149. *monticola* Sturm
56(54) Maxillary palpi and legs rufo-testaceous. Elytra over their
entire surface with transverse microsculpture without evi-
dent meshes . . . . 154. *stephensi* Crotch
  – Penultimate segment of maxillary palpi dark. Legs infu-
scated. Microsculpture of elytra indistinct in basal half, or
with evident meshes (at least near apex) 57
57(56) Microsculpture of elytra consisting of isodiametric meshes
(or almost so), in the ♂ sometimes visible only near the
apex. Antennae almost moniliform, that is, with outer seg-
ments rounded (Fig. 231) . . . . 153. *dauricum* (Motschulsky)
  – Microsculpture consisting of transverse lines or transverse
meshes. Outer antennal segment more stretched and coni-
cal (Figs 229, 230) 58
58(57) Microsculpture of elytra (even at apex) consisting of trans-

verse lines. Antennae with shorter outer segments. Penis,
Fig. 223 .... 151. *grapii* Gyllenhal
- Elytral microsculpture forming evident transverse meshes
(in the ♂ only near apex) 59
59(58) Black, upper surface vividly bluish, green or almost blue.
Legs rufo-testaceous with piceous femora. Elytral
striae with coarse punctures. Wings full. Penis, Fig. 221 ....
.... 150. *nitidulum* (Marsham)
- Piceous to almost black with greenish or aeneous lustre
(very rarely vivid as in *nitidulum*). Legs piceo-ferrugineous
with slightly infuscated femora. Elytral striae with moderate punctures. Wings usually rudimentary. Penis, Fig. 222 60
.... 152. *yukonum* Fall
60(59) Elytra only with large, arcuate apical lunula, base not maculate................................... 155. *lunatum* (Duftschmid)
- Elytra with both basal and preapical pale spot, usually se-

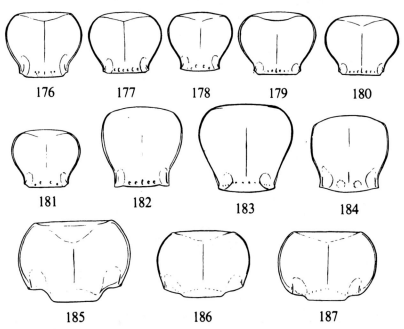

Figs 176-187. Pronotum of *Bembidion*. — 176: *minimum* (F.); 177: *normannum* Dej.; 178: *azurescens* (D. Torre); 179: *quadrimaculatum* (L.); 180: *humerale* Sturm; 181: *tenellum* Er.; 182: *articulatum* (Pz.); 183: *genei* Küst.; 184: *doris* (Pz.); 185: *aeneum* Germ.; 186: *guttula* (F.); 187: *mannerheimii* Sahlb.

parated by a transverse dark fascia (sometimes confluent; often indistinct in *obscurellum*) 61

61(60) Elytra dull from dense reticulate microsculpture consisting of isodiametric or slightly transverse meshes. Antennae quite pale or slightly infuscated towards apex. Dark colour of elytra often reduced 62
   – Elytra somewhat shiny, microsculpture consisting of transverse lines, often joining into strongly transverse meshes. At most 3 antennal segments pale 63

62(61) 5.0-5.5 mm. Pale elytral spots coherent along side margin ....
   .... 161. *maritimum* (Stephens)
   – 3.0-5.1 mm. Pale elytral spots normally leaving only 1st interval, extreme side-margin and a triangular spot across the suture behind middle, dark .... 162. *obscurellum* (Motschulsky)

63(61) Pronotum micro-reticulate over its entire surface and therefore not shiny .... 157. *bruxellense* Wesmael
   – Pronotum, at least on disc, without microsculpture and very shiny 64

64(63) 7th elytral stria evident anteriorly as a row of well-defined

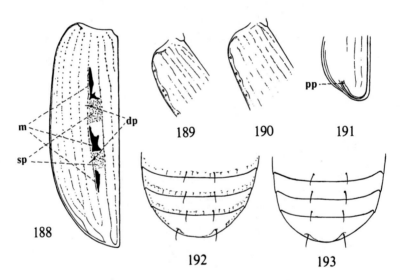

Figs 188-193. *Bembidion.* — 188: left elytron of subg. *Chrysobracteon,* dp = dorsal punctures, m = "mirror", sp = "silver spot". — 189, 190: basis of left elytron of 189: subg. *Plataphodes,* with angulate shoulder, and 190: subg. *Plataphus* with rounded shoulder. — 191: elytron of subg. *Ocys,* with recurrent sutural stria, pp = preapical puncture. — 192, 193: abdominal sternites of 192: *virens* Gyll. and 193: *prasinum* (Dft.).

punctures. Base of pronotum with rather coarse punctuation. Wings usually reduced (not longer than elytra) ... 156. *tetracolum* Say
- 7th stria rudimentary (as a row of faint, irregular punctures) or wanting. Base of pronotum with weak, confluent punctuation, or absent 65
65(64) Elytra with faint metallic, usually bronzy, hue; 7th stria rudimentary but evident; elytral microsculpture forming transverse meshes .... 158. *petrosum* Gebler
- Elytra without or with extremely faint bluish hue, 7th stria virtually disappeared; elytral microsculpture transverse with clear formation of meshes 66
66(65) Legs and 3 basal segments of antennae entirely pale 159. *andreae* (Fabricius)
- 3rd antennal segment and femora clearly infuscated 160. *femoratum* Sturm
67(8) Pronotum (Fig. 163) with base straight. Shoulders angulate with base margined laterally .... 101. *obtusum* Audinet-Serville
- Base of pronotum sinuate laterally (Figs. 185, 186). Shoulders rounded 68
68(67) Lateral sinuation of pronotal base deep (Fig. 185). 3.4-5.5 mm 69
- Lateral sinuation of pronotal base shallow (Fig. 186). 2.8-3.5 mm 72
69(68) 7th elytral stria well developed with coarse punctures anteriorly. Upper surface with strong blue-green reflection, elytra strongly iridescent .... 131. *biguttatum* (Fabricius)
- 7th stria absent or faintly suggested. Upper surface at most with moderate metallic hue, elytra somewhat iridescent 70
70(69) Pronotum densely microsculptured over its entire surface. Elytra with fine striae, intervals flat .... 134. *aeneum* Germar
- Pronotum with disc smooth and shiny. Elytral striae strongly punctate and intervals convex 71
71(70) Antennae very slender, segments 8-10 more than twice as long as wide. Elytral striae less strongly punctate (as in *biguttatum*) .... 132. *iricolor* Bedel
- Antennae stouter, segments 8-10 less than twice as long as wide. Punctuation of elytral striae stronger than in *biguttatum* .... 133. *lunulatum* (Fourcroy)
72(68) Legs and 1st antennal segment bright rufous. Pronotum (Fig. 187) broader, with more rounded sides. Upper surface without or with faint bluish hue .... 136. *mannerheimii* Sahlberg
- Legs and 1st antennal segment more brownish. Pronotum (Fig. 186) with more defined hind-angles. Upper surface with more evident, blue or green reflection .... 135. *guttula* (Fabricius)

---

[1] Note. According to Andersen (1970) females may have three to six setigerous punctures on the last, sometimes also on the last but one, segment.

Subgenus *Bracteon* Bedel, 1879

Bracteon Bedel, 1879, Faune Col. Bass. Seine 1: 27.
Type-species: *Carabus striatus* Fabricius, 1792.

Closely related to the following subgenus but without the special structures of 3rd interval. Agreeing through the angulate shoulders, the 9 complete, punctate elytral striae and the strong microsculpture. The behaviour is also the same.

92. *Bembidion striatum* (Fabricius, 1792)

Elaphrus striatus Fabricius, 1792, Ent. Syst. 1: 179.

5.5-6.5 mm. General outline as *B. velox*. Under surface green, upper surface with a bronze lustre, dull because of strong reticulate microsculpture; tibiae, base of femora and at least 1st antennal segment rufo-testaceous.

Distribution. Denmark: only one accidental specimen recorded from LFM, Ulfshale, 1932. It was found on the sea shore. No records from Sweden or Norway. — Finland: found in drift material in N: Ekenäs, Klovaskär, 1939 (2 specimens). — USSR: a few specimens from the Svir area. — From W.Europe to Siberia.

Biology. On sandy river banks, sometimes together with *B. velox* or *litorale*. In Finland and Denmark only accidental findings on the seashore.

Subgenus *Chrysobracteon* Netolitzky, 1914

Chrysobracteon Netolitzky, 1914, Ent. Blätt. 10: 166.
(Other names given by Netolitzky 1940, e.g. *Litorebracteon* and *Argyrobracteon,* are superfluous).
Type-species: *Carabus velox* Linnaeus, 1761.

Third elytral interval dilated, with two dull "silver spots" carrying a dorsal puncture and in front of each with a dark shiny "mirror" (Fig. 188). All other intervals (except the outer in *litorale*) equally microsculptured.

All species occur on open sand near water and are extremely agile, running and flying in bright sunshine.

93. *Bembidion velox* (Linnaeus, 1761)
Figs 158, 194.

Carabus velox Linnaeus, 1761, Fauna Suec. ed. 2: 222.
Bembidion Güntheri Seidlitz, 1887, Fauna Baltica ed. 2: 64.

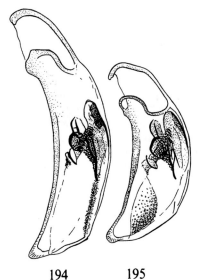

194   195

Figs 194, 195. Penis of 194: *Bembidion velox*
(L.) and 195: *B. lapponicum* Zett.

5.0-6.0 mm. Blue-green below, upper surface vividly metallic, usually brassy or coppery, with black "mirrors" and often greenish "silver spots". At least 1st antennal segment and legs, except knees and tarsi, bright rufous. A form with dark blue or blue-
green colour and more or less metallic appendages was called *"guentheri"*. Pronotum
(Fig. 158) with strongly rounded sides. Penis, Fig. 194.

Distribution. Not in Denmark. — Sweden: rather distributed throughout the country but completely lacking in the eastern part of S.Sweden (except a few finds on Gotland and Gotska Sandön). More or less common in central and northern Sweden, rare
(or apparently extinct) in the south (almost all records are before 1950). — Norway:
fairly common in the south and south-east, rare in central and northern districts. No
records from VE, HO, SF and MR. — Finland: fairly common over most of the country, lacking mainly in the alpine zone. — In the Soviet part of E.Fennoscandia north
to Pechenga. — N. and C.Europe, Siberia.

Biology. On barren sand on river and lake shores, less often by the sea, occurring on
firm moist sand close to the water edge. It is a heat-loving beetle, running and flying
about in sunny weather. During inactive periods it is buried in the sand or hidden under debris, etc. Most numerous in May-July. The beetles regularly dig burrows in the
sand, and Andersen (1978) assumes that females, like those of *B. argenteolum,* deposit
their eggs at the bottom of such burrows.

143

94. **Bembidion lapponicum** Zetterstedt, 1828
  Figs 157, 195.

Bembidium lapponicum Zetterstedt, 1828, Fauna Ins. Lapp. 1: 6.
Bembidium latiusculum Motschulsky, 1844, Ins. Sibér.: 272.
Bembidion jenisseense J. Sahlberg, 1880, K. Svenska Vetensk Akad. Handl. 17: 12, 14.

4.4-5.9 mm. Closely related to *velox* and earlier regarded as a subspecies; but the penis is quite different. Upper surface dull, with a greyish green (sometimes bluish or brassy) lustre. Only 1st antennal segment (as a rule only underneath) and of the legs at most tibiae and the base of femora dark rufous. Pronotum (Fig. 157) with less rounded sides. 3rd interval of elytra more suddenly dilated in front of anterior "silver spot" and on the whole broader as compared with 4th interval. The penis (Fig. 195) is much shorter and more arcuate. Internal sac with a very characteristic ventral, scaly "pillow".

Distribution. Not in Denmark. — Sweden: rare in the western mountains from Härjedalen to Lapland, generally distributed only in the extreme north. — Norway: rather scattered distribution from ST (63°N) and northwards. — Finland: northernmost parts, south to Pallastunturi and Kuusamo. — USSR: Lr and northern Kr. — A circumpolar species.

Biology. In the birch and upper conifer region on sparsely vegetated or barren shores of lakes and rivers, often on alluvial soil in deltas of brooks. Preferably on weakly moist sand, less often on gravelly or silty sites. The beetles are very active in warm, sunny weather, often taking away when disturbed. They may shelter in rather dense vegetation. The species digs burrows in the sand and probably has a breeding behaviour similar to that of *B. argenteolum* (Andersen 1978). Reproduction takes place mainly in June-July.

95. **Bembidion litorale** (Olivier, 1791)
  Fig. 159; pl. 5: 6.

Elaphrus litoralis Olivier, 1791, Encycl. Meth. 6: 353.
Elaphrus paludosus Panzer, 1794, Fauna Ins. Germ. 20: 4.

5.2-6.2 mm. Colour more variegated than in related species, elytra usually coppery or purplish and grey; also outer intervals with opaque spots, 5th and 7th usually with extra "mirrors". Bluish or nearly black specimens are rare. Appendages, except underside of 1st antennal segment, dark. Pronotum (Fig. 159) narrow and convex, hindangles little protruding. 4th elytral stria suddenly bent in front of anterior "silver spot".

Distribution. Denmark: very rare, recorded from riverine habitats in SJ, EJ and WJ, and from LFM: in great numbers 10.vi.1984 in a clay-pit at Ny Borre, Høvblege on Møn. — Sweden: rare and extremely local in Sm., Vg., Vrm., Hls., Med., Ång., Vb.,

Nb. and Ås. Lpm. Only old records south of Vrm. — Norway: rare and local, recorded from two separate areas: in the south-east (HE, B) mainly along the river Glomma; in ST and NT scattered, but locally common. — Finland: fairly rare but widely distributed in the inner part of the country. — USSR: rather common around Ladoga and Svir, recorded also from northern Kr. — Europe, Siberia.

Biology. In Scandinavia confined to river banks, in C.Europe also found on the shores of standing waters, mainly occurring on sun-exposed, sparsely vegetated silty sand, less often on pure sand. It is less hygrophilous than the other *Chrysobracteon*-species, living at some distance from the water, even on completely dry soil. The beetles are mostly active in warm sunshine; under unfavourable weather conditions they often shelter in denser vegetation. Flying occurs less often than in the related species. The reproductive period is in early spring, predominantly in May.

96. *Bembidion argenteolum* Ahrens, 1812
    Fig. 160.

Bembidion argenteolum, Ahrens, 1812, Neue Schr. d. Naturf. Ges. Halle 2(2): 23.

5.9-7.5 mm. The largest species of the entire genus. Upper surface brassy, often with a greenish, rarely bluish, hue. 1st antennal segment, tibiae and base of femora more or less pale. Elytra less microsculptured and therefore more shiny than in preceding species. Pronotum, Fig. 160. 3rd elytral interval broader than in *velox*.

Distribution. Not in Denmark. — Sweden: restricted to a small area in central Sweden (Värmland, Dalarna, Hälsingland) but rare and local. In earlier times often collected at the river Klarälven, in recent time with strongly decreasing frequency. — Norway: rare and local, the distribution being nearly identical to that of the preceding species (HE, O, B, ST, NT). — Finland: has been listed on the basis of specimens labelled "Suomussalmi, Sorsakoski". It has previously been discovered that some specimens labelled so were incorrectly labelled, and this seems to be another case. — USSR: collected in large numbers on the Karelian Isthmus; also found in the Svir area. — N. and C.Europe, Siberia.

Biology. Habitat as for *B. velox*, i.e. sandy shores usually lacking vegetation, but mostly at some distance from the water edge on rather dry, fine sand. Predominantly along rivers, rarely on lake and sea shores. The beetles are active on the surface in warm sunshine and often flies when disturbed. When inactive, they usually stay in burrows in the sand. Breeding occurs mainly in June. According to Andersen (1966, 1978) eggs are deposited separately in small egg-chambers at the end of a burrow dug by the female.

Subgenus *Neja* Motschulsky, 1864

Neja Motschulsky, 1864, Bull. Soc. Nat. Moscou 37(2): 188.
Type-species: *Bembidion ambiguum* Dejean, 1831.

The single species has the frontal furrows doubled behind the eyes. Shoulders angulate.

97. *Bembidion nigricorne* Gyllenhal, 1827
Fig. 161.

Bembidium nigricorne Gyllenhal, 1827, Ins. Suec. 4: 402.

3.4-3.8 mm. Similar to *lampros* in size and colour, but antennae entirely black; tibiae sometimes brown. Pronotum (Fig. 161) very little sinuate before the non-protruding hind-angles. Elytra with rows of coarse punctures, barely joined by a stria. The species shows wing-dimorphism.

Distribution. Very rare in Denmark, only a few localities in the central parts of Jutland (districts EJ, WJ, NEJ). — Sweden: local and regarded as being very rare but probably rather widely distributed. As yet recorded for Hall., Sm., Ög., Vg., Nrk., Vrm., Dlr., Hls., Hrj. and Nb. — Norway: very rare, recorded for the first time from this country in 1979, at Sørumsand, AK. — Finland: very rare but widely distributed, recorded from Ab, N, St, Ta, Sa, Oa, Om, Ok and ObS. — USSR: a few records from Vib and Kr. — N. and northern C.Europe.

Biology. On *Calluna*-heaths, on dry sandy ground (often together with *Amara infima*), or on more or less humid peaty soil (sometimes in company with *B. humerale*). It is most numerous on sparsely vegetated spots, especially on regenerating *Calluna*-areas. *B. nigricorne* is a winter breeder, mainly occurring in October-December (Grossecappenberg *et al.* 1978).

Subgenus *Chlorodium* Motschulsky, 1864

Chlorodium Motschulsky, 1864, Bull. Soc. Nat. Moscou 37(2): 182.
Type-species: *Bembidion colchicum* Chaudoir, 1850.

Easily recognized on the dull upper surface, caused by dense reticulate microsculpture. Shoulders angulate.

98. *Bembidion pygmaeum* (Fabricius, 1792)

Carabus pygmaeus Fabricius, 1792, Ent. Syst. 1: 167.

3.5-4.2 mm. Superficially similar to *lampros* but pronotum much broader with greatest width about middle; elytra more parallel-sided, the striae extremely finely punctate. Metallic lustre bronzy. Elytra often with pale subapical spot. Antennal base underneath and tibiae more or less pale.

Distribution. Not in Denmark, Sweden or Norway. — Finland: only found once in sea drift in N: Tvärminne (1939). — USSR: Vib and southern Kr, north to Karhumäki. — Europe from the Pyrenees to NW.Russia.

Biology. In E. Fennoscandia confined to river banks, occurring on more or less dry, sandy or clayey spots with sparse vegetation, not close to water. In C.Europe often in clay pits.

Subgenus *Metallina* Motschulsky, 1850

Metallina Motschulsky, 1850, Käf. Russl.: 13.
Type-species *Bembidion lampros* (Herbst, 1784).

Upper surface entirely without microsculpture, very shiny. Pronotum with a group of small whitish bristles at front-angles. Frontal furrows deep and simple (Figs 149, 150). Shoulders angulate. Both species are wing-dimorphic.

99. *Bembidon lampros* (Herbst, 1784).
Figs 149, 196, 197.

Carabus lampros Herbst, 1784, Arch. Insectengesch. 5: 143.
Carabus celer Fabricius, 1792, Ent. Syst. 1: 167.

3.0-4.4 mm. Upper surface with a metallic, usually brassy, rarely bluish, lustre. Base of antennae (at least 1st segment underneath) and legs rufous, but femora and tarsi often infuscated. Frontal furrows (Fig. 149) somewhat dilated at middle. Pronotum pronouncedly cordiform. Elytra with rounded sides, 7th stria lacking or consisting of a row of weak, irregular punctures. Penis (Fig. 197) with an external leftside fold; internal sac with a complicated armature.

Distribution. Denmark: very common and generally distributed, known from all districts. — Sweden: very common and generally distributed over the entire country except in the extreme north (no records from T. Lpm.). — Norway: very common south of 64°N (NT). Only one record from Ns, not found in TR and F. — Finland: found in most of the country (not Le and Li), very common in the south. — USSR: common in the southern parts, not recorded from Lr. — Europe and Siberia; introduced in North America.

Biology. Very eurytopic, living on almost every kind of open, sun-exposed ground with sparse vegetation, often on dry sandy soil. Frequently on cultivated land, notably in root crop fields. The adult feeds mainly on small arthropods, also on insect eggs, and may play an important role in the natural control of e.g. the cabbage root-fly. It is a typical spring breeder.

147

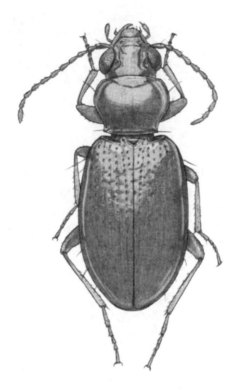

Fig. 196. *Bembidion lampros* (Hbst.), length 3-4.4 mm.

100. *Bembidion properans* (Stephens, 1828)
Figs 150, 198.

Tachypus properans Stephens, 1828, Ill. Brit. Ins. Mand. 2: 26.
Bembidion velox Erichson, 1837, Käf. Mark Brandenburg 1(1): 134; *nec* (Linnaeus, 1761).
Bembidion quatuordecimstriatum Thomson, 1871, Opusc. Ent. 4: 361.

3.5-4.2 mm. Long regarded as a form or subspecies of *lampros,* but has a quite different penis. Somewhat flatter, elytra more parallel-sided. Frontal furrows (Fig. 150) not dilated at middle. Sides of pronotum more broadly depressed. 7th stria consisting of evident, regular punctures, at least anteriorly. Penis (Fig. 198) without lateral fold; internal sac with much more developed sclerites.

148

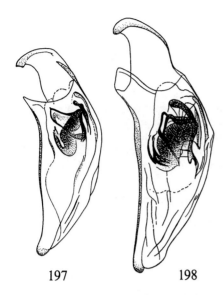

197   **198**

Figs 197, 198. Penis of 197: *Bembidion lampros* (Hbst.) and 198: *B. properans* Stph.

Distribution. Denmark: rare in Jutland, most localities are in EJ; rather distributed and common on the islands. — Sweden: from Skåne to Dalarna, usually rather rare and somewhat local. — Norway: rather common in the south-east (Ø, AK, HE), moreover a few records from northern parts of O. — Finland: rather common in the south, north to Kuopio in Sb. — Also widely distributed in the southern Soviet part of E.Fennoscandia. — Europe and Siberia.

Biology. On sun-exposed ground with sparse vegetation, usually on moderately humid, clayey or clay-mixed soil, never on dry sandy sites. Often in clay and gravel pits. Mainly in spring and early summer.

Subgenus *Phyla* Motschulsky, 1844

Phyla Motschulsky, 1844, Ins. Sibér.: 260.
Phila auctt.
  Type-species. *Bembidion obtusum* Audinet-Serville, 1821.

Pronotum (Fig. 163) without latero-basal sinuation and with a straight base. Shoulders angulate.

101. *Bembidion obtusum* Audinet-Serville, 1821
  Fig. 163.

Bembidion obtusum Audinet-Serville, 1821, Faune Franç. 1: 83.

2.8-3.5 mm. Piceous brown, sometimes with a bluish or greenish hue, elytra faintly iridescent, pronotum and elytral suture often paler. Base of antennae and legs rufous, femora usually infuscated. Pronotum without, elytra with dense, transverse microsculpture. 6th elytral stria suggested anteriorly only, 7th stria disappeared. Wings either full or completely reduced.

Distribution. Denmark: well distributed in eastern Jutland and on the islands, sparsely occurring in or absent from many areas in western and northern Jutland. — Sweden: a southern species, distributed in the lowlands and coastal districts north to Ög. (several localities after 1950), Vg. and Nrk. (3 localities after 1980). In western Skåne and on Öland and Gotland often common, elsewhere more sparse. — Not recorded from Norway or East Fennoscandia. — Widely distributed in C.Europe; also known from Madeira.

Biology. On exposed or faintly shaded, moderately dry clay-soil. It is mainly associated with agricultural land, e.g. occurring in potato and cereal fields, less often in shore habitats and on clayey slopes. A spring breeder.

Subgenus *Ocys* Stephens, 1828

Ocys Stephens, 1828, Ill. Brit. Ins. Mand. 2: 2.
Type-species: *Bembidion harpaloides* Audinet-Serville, 1821.

This is an isolated group, with sutural stria of elytra "recurrent" (as in *Trechus*, Fig. 191) and externally delimited by a keel. Dorsal punctures (1 or 2) situated behind middle. Shoulder not or barely angulate. Wings full.

102. *Bembidion harpaloides* Audinet-Serville, 1821
   Fig. 164.

Bembidion harpaloides Audinet-Serville, 1821, Faune Franç. 1: 78.
Tachys rufescens Guerin, 1823, Bull. Sci. Philom. Paris: 123.

4.2-6.0 mm. Unique within the entire genus by the presence of only one dorsal elytral puncture. Rufous, elytra darker, at least apically, sometimes with a bluish hue. Pronotum (Fig. 164) with a straight base and sharp hind-angles. Elytra with a suggested humeral angle. Outer striae obliterated.

Distribution. In our area only found in Norway: two records from VA (Mandal and Lyngdal). — Western Europe, including the Azores.

Biology. Usually found under bark, at the base of trees, or under stones, on clayish, rather moist ground, often along rivers. It may be associated with animals' nests; immature adults have been observed in the nest of a jay in England.

150

103. *Bembidion quinquestriatum* Gyllenhal, 1810
    Fig. 165.

Bembidium quinquestriatum Gyllenhal, 1810, Ins. Suec. 2: 34.

3.5-4.3 mm. Superficially similar to *Trechus quadristriatus,* also in the form of pronotum (Fig. 165), but elytral striae evidently punctate. Piceous or reddish brown, usually
with a bluish or greenish hue; appendages pale. Pronotum with hind-angles obtuse
and base oblique laterally. 5th elytral stria weak, 6th and 7th more or less disappeared;
shoulder completely rounded; 2 dorsal punctures behind middle.

Distribution. Denmark: recorded from only a few localities in SJ, WJ, EJ, NEJ,
LFM, SZ, NEZ and B; very rare and usually singly. — Sweden: scattered distributed
(Sk. — Upl.) and extremely rare. In the 19th century found on several localities, in the
last 40 years only the following localities: Sk., Ålabodarna (1955); Hall., Varberg
(1970); Vg., Göteborg; Öl., Borgholm (1969 and later). — Norway: very rare, one old
record from AK, later also found in VE. — Not found in East Fennoscandia. — Most
of Europe except the north and south-west.

Biology. Pronouncedly synanthropic, occurring in and around cellars, stables,
ruins, etc., mostly found crawling on walls; sometimes under bark of trees in gardens.
It is most probably associated with rodents. The adults become active already in
February-March and are presumably winter breeders (Baranowski 1977).

Subgenus *Princidium* Motschulsky, 1864

Princidium Motschulsky, 1864, Bull. Soc. Nat. Moscou 37(2): 181.
    Type-species: *Bembidion punctulatum* Drapiez, 1820.

This and all following subgenera have rounded, not angulated elytral shoulders. Forebody coarsely punctate; frontal furrows only suggested. Elytral striae complete to
apex. Wings full.

104. *Bembidion punctulatum* Drapiez, 1820

Bembidion punctulatum Drapiez, 1820, Annls Gen. Sci. Phys. Brux. 7: 275.

4.5-5.6 mm. Similar to *bipunctatum,* but also disc of pronotum punctate, elytra broader with strongly punctate striae, convex intervals, and a flat transverse impression in
the anterior half. Black, upper surface with a strong, usually bronze lustre. 1st antennal segment and legs pale.

Distribution. Not in Denmark, Sweden or Norway. — Finland: recorded once from
Ka: Hamina, probably accidental. — USSR: common in the Karelian Isthmus
(southern part), altso recorded in southern Kr. — Europe, North Africa and the Caucasus.

Biology. Usually on river banks, predominantly on barren, stony or gravelly sites, together with e.g. *B. saxatile* and *bipunctatum*. Also found on sandy shores in company with *B. litorale*. Rarely at the border of still water.

Subgenus *Testedium* Motschulsky, 1864

Testedium Motschulsky, 1864, Bull. Soc. Nat. Moscou 37(2): 182.
Type-species: *Carabus bipunctatus* Linnaeus, 1761.

Head densely, coarsely punctate as in preceding subgenus. Elytral striae obliterating towards apex; dorsal punctures foveate. Wings full.

105. *Bembidion bipunctatum* (Linnaeus, 1761)
Pl. 5: 20.

Carabus bipunctatus Linnaeus, 1761, Fauna. Suec. ed. 2: 223.

3.6-4.7 mm. At once recognized on the elytral foveae. Upper surface metallic, usually brassey, often greenish, rarely blue or even entirely black; occasionally with rufinistic elytra. Appendages black. Pronotum impunctate at middle. Elytral intervals flat.

Distribution. Denmark: only few localities, usually near the coast; recorded from all districts except F and SZ; may be termed very rare. — Sweden: a widespread but usually not common species, known from all districts except G. Sand. and Nrk. Most frequent along the coasts. — Norway: widely distributed and common in most districts, particularly in the north. Scattered and local in the south-east. — Finland: found all over the country and is mostly common. — USSR: all over Vib, Kr and Lr. — Europe, North Africa and western Siberia.

Biology. At rivers, lakes and on sea shores on different kinds of soil, sometimes far from open water. Often on barren river banks among stony gravel in company with *B. saxatile,* in the north also with *prasinum* and *virens.* On silty sites some vegetation of e.g. *Carex* or grass seems to be required. It is often common in the birch region of the mountains. In Denmark the species is most predominant on clayey salt marsh ground. Mainly in spring and early summer. Most individuals hibernate as adults, in N.Fennoscandia some probably as larvae.

Subgenus *Paraprincidium* Netolitzky, 1914

Paraprincidium Netolitzky, 1914, Ent. Blätt. 10: 165.
Type-species: *Elaphrus ruficollis* Panzer, 1797.

A single pale small species with short stature. Punctuation of head as in the two preceding subgenera. Elytral striae sharp to apex. Wings full.

106. *Bembidion ruficolle* (Panzer, 1797)

Elaphrus ruficollis Panzer, 1797, Fauna Ins. Germ. 38: 12.

3.3-3.5 mm. Rufo-testaceous (head darkest) with a greenish lustre, especially on head, base of pronotum and in the elytral striae; a dark cloud on the centre of elytra. Pronotum impunctate on disc. Superficially similar to the following species but smaller, with complete elytral striae.

Distribution. Only 3 accidental finds in Denmark of stray individuals. — Sweden: very rare with scattered distribution in the southern half of the country (Sk., Sm., Gtl., G. Sand., Vrm., Dlr. and Med.). Only a few localities known, most of them in central and eastern Skåne but usually abundant where occurring. — Not in Norway. — Finland: rare but widely distributed, north to LkE. — USSR: many records from the Karelian Isthmus; otherwise scattered finds north to northernmost Kr. — From C.Europe to W.Siberia.

Biology. Predominantly on river banks, less often on lake shores, occurring on barren or sparsely vegetated sandy sites, close to the water edge. The adults are active on the surface in warm sunshine; otherwise buried in the sand.

Subgenus *Actedium* Motschulsky, 1864

Actedium Motschulsky, 1864, Bull. Soc. Nat. Moscou 37(2): 182.
Type-species: *Elaphrus pallidipennis* Illiger, 1802.

A short species with narrow forebody. Colour variegated. Punctuation of head as in the three preceding subgenera. Elytral striae obliterating towards apex. Wings full.

107. *Bembidion pallidipenne* (Illiger, 1802)
Pl. 5: 7.

Elaphrus pallidipennis Illiger, 1802, Magazin Insektenk. 1: 489.

4.1-4.7 mm. Lower surface, head and pronotum black with a metallic, usually greenish lustre; elytra pale yellow with a spot around scutellum and an irregular transverse fascia behind middle piceous. Appendages pale. Head a little narrower than elytra, twice as broad as pronotum.

Distribution. Denmark: found along the coasts throughout the country; also a few inland finds. — Sweden: along the coasts from Göteborg (Vg.) to Öl., Gtl. and G. Sand., usually rather rare. In Skåne also found on a few inland localities (Ivösjön, Ringsjön, Herrevadskloster, Vittsjö and Nöbbelöv near Tollarp). — Norway: very rare and local. One record from AK, scattered along the coast of VA and R. — Not in Finland or in the Soviet part of E.Fennoscandia. — W.Europe.

Biology. A halophilic species, predominantly occurring on seashores on moist, fine and rather loose sand, either on barren sites or among sparse vegetation of e.g. *Puccinellia* or *Atriplex*. Rarer on lake shores inland. On beaches often together with *Bledius fergussoni* Joy (*arenarius* Payk.) and *Heterocerus hispidulus* Kies., both of which are preyed upon by *B. pallidipenne*, and with *Dyschirius obscurus* and *thoracicus*. The beetles are nocturnal; during daytime they stay in burrows in the sand. Reproduction takes place in spring. Eggs are laid in batches in burrows dug by the female (Larsen 1936).

Subgenus *Eupetedromus* Netolitzky, 1911

Eupetedromus Netolitzky, 1911, Wien. Ent. Ztg. 30: 190.
Type-species: *Carabus dentellus* Thunberg, 1787.

This and the three following subgenera are distinguished by the dorsal elytral punctures of which at least the anterior is "free", that is, placed on 3rd interval without touching any of the adjacent striae. The present subgenus contains large species with elytra iridescent from dense, transverse microsculpture. The anterior supra-orbital puncture is surrounded by an elevated, shiny field. Elytral striae complete. Wings full.

108. *Bembidion dentellum* (Thunberg, 1787)
Figs 166, 199.

Carabus dentellus Thunberg, 1787, Mus. Nat. Ac. Ups. Diss. Upsala 4: 50.

5.1-6.0 mm. Black, forebody with a bronze lustre, elytra with variegated pattern, consisting of many small rectangular pale spots, confluent basally and fusing into an irregular transverse band behind middle. Antennal base (at least 1st segment) and legs pale. Pronotum, Fig. 166. Elytra near base with a shallow, transverse impression; shoulder-tooth rudimentary. Penis, Fig. 199.

Distribution. Denmark: stable populations only found on Bornholm. Stray individuals found 1976 and 1981 on S. Falster (Bøtø). — Sweden: generally distributed and rather common from Sk. to Ång. (very rare in S.Skåne). Completely absent from the western part of the country from N.Dalarna and towards the north. A doubtful record from Nb. — Norway: rather common in the south-eastern districts; otherwise mainly single records from HO, SF, MR, ST and NT. — Finland: fairly common in the south (not known from Al), recorded north to Om: Kalajoki and Oulainen; one specimen from LkW (?mislabelled). — USSR: Vib, Kr and southern part of Kola peninsula. — Europe and Siberia.

Biology. On soft muddy or clayey, somewhat shaded ground on the shores of standing or slowly running waters. Usually in rather dense vegetation of *Juncus,* grasses, *Equisetum*, etc., often on mossy carpets; also under *Salix* or *Alnus* among dead leaves and twigs. A spring breeder, which now and then hibernates also as larva or pupa (Andersen 1970).

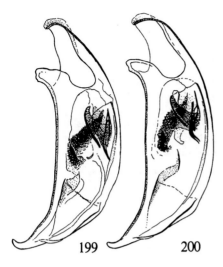

Figs 199, 200. Penis of 199: *Bembidion dentellum* (Thbg.) and 200: *B. tinctum* Zett.

109. *Bembidion tinctum* Zetterstedt, 1828
Fig. 200.

Bembidion tinctum Zetterstedt, 1828, Ins. Lapp. 1: 8.

5.0-6.0 mm. Closely related to *dentellum* but somewhat slenderer with more parallel-sided elytra. Upper surface with a more piceous ground-colour and the pale spots less bright, causing less pronounced colour contrasts. Pronotum with broader base. Elytra with more evident shoulder-tooth; their striae usually shallower and more weakly punctate. Penis (Fig. 200) with apex slenderer, less arcuate.

Distribution. Not in Denmark or Norway. — Sweden: scattered distributed in the northeast, south to Med. Rather common in Nb. — Finland: in the northern part, south to ObS: Pudasjärvi and Kb: Nurmes; also found in St: Noormarkku and Oa: Kauhajoki. — USSR: Lr and northern Kr. — East to Siberia.

Biology. Like the preceding species on soft muddy or clayey shores of rivers and still waters, usually in less shady situations. Often among vegetation of *Carex,* etc. Mainly in June.

Subgenus *Notaphus* Dejean, 1821

Notaphus Dejean, 1821, Cat. Coll. Col. B. Dejean: 16.
Type-species: *Carabus varius* Olivier, 1795.

Small to moderate species with a characteristic "mozaic" elytral pattern, usually confluent into two transverse fasciae. Their microsculptue coarse and more or less reticulate. Frons without elevated field laterally and with shallow furrows. Wings full.

155

110. *Bembidion varium* (Olivier, 1795)
Fig. 167; pl. 5: 8.

Carabus varius Olivier, 1795, Ent. 3(35): 110.
Bembidium ustulatum Sturm, 1825, Deutschl. Fauna Ins. 5(6): 158.

4.1-5.1 mm. Largest species of the group, but smaller than *dentellum*. Black with a bronze lustre, often somewhat greenish (rarely bluish). The pale elytral spots sometimes diffuse, confluent anteriorly, or reduced without forming transverse bands. At least 1st antennal segment (often 2nd to 4th underneath) and legs brown, though femora with metallic hue. Pronotum, Fig. 167. Elytral striae fine but punctate, intervals flat. Microsculpture of elytra regular, with meshes arranged as bricks.

Distribution. Denmark: rather common and distributed, recorded from all districts. — Sweden: in the southeast, generally distributed and rather common in Sk., Öl. and Gtl.; only one or a few localities in Bl., Hall., Sm., G. Sand., Ög., Vg. and Sdm., north to Upl. and Ång. (Öre älv, Lillnäset, 1984). — Norway: very rare, recorded for the first time in 1983, at Eidsberg, Ø. — Finland: rare, found in the south-western part, north to Tb: Konnevesi. — Not in Soviet part of E.Fennoscandia. — From Europe and North Africa to Siberia and Mongolia.

Biology. On moist sun-exposed clay with patchy vegetation of *Carex*, grasses, etc., notably in salt marshes, but also at the margin of fresh water. The beetles are active on the surface and frequently flies in warm sunshine. They are often found in wind drift. The adults prey upon small worms, insect larvae (e.g. of *Heterocerus* and *Ochthebius*), etc. Breeding occurs in spring.

111. *Bembidion semipunctatum* (Donovan, 1806)

Carabus semipunctatus Donovan, 1806, N.H. Brit. Ins. 9: 22.
Bembidion adustum Schaum, 1860, Naturg. Ins. Deutschl. Col. 1(1): 686.

3.2-4.0 mm. Smaller and shorter than *varium*. The metallic lustre usually more yellowish (very rarely bluish), pale elytral spots always sharp. Legs paler, 3 or 4 segments of antennae entirely pale (or with faintest metallic hue above). Base of pronotum somewhat broader. Elytral striae deep, coarsely punctate, intervals more or less convex; microsculpture more irregular.

Distribution. Denmark: only a few captures on Bornholm, probably of stray individuals. — Sweden: very restricted distribution, partly along the rivers Klarälven (Vrm.) and Dalälven (Dlr.), partly in the southeastern part of the country (Sk., Gtl.). The finds in the latter area are probably accidental, and a result of migrating specimens (5 localities). At river Klarälven rather common (at least earlier). — Norway: scattered but very local in the eastern part (AK, HE, O, B, VE) north to ST and NT. — Finland: several specimens found in sea drift in N: Tvärminne, 1939. — USSR: many records from the Karelian Isthmus; also found north of Ladoga. — From Europe and North Africa to western N.America.

Biology. In Scandinavia almost confined to river banks, in C.Europe also at the margins of standing waters. Preferably in silty, moderately humid and somewhat shady sites among grass, *Juncus, Equisetum,* etc., often on mossy ground, frequently under bushes. The beetles often run about in sunny weather on exposed spots. Sometimes encountered in wind drift on the seashore. Breeding occurs in spring.

112. *Bembidion obliquum* Sturm, 1825

Bembidium obliquum Sturm, 1825, Deutschl. Fauna Ins. 5(6): 160.

3.0-4.4 mm. In normal coloration much darker than the two preceding: black, upper surface with rather weak brass or greenish metallic lustre (sometimes blue). Lower surface of 1st antennal segment rufo-piceous, legs almost black; elytral pale pattern very distinct, besides the two fascia with only few small spots, apex and epipleura (against the two preceding) black. However, the species is much varying in colour. The darkest specimens are seemingly unicolorous (though elytral spots are revealed in transparent light). The palest specimens may be coloured as *semipunctatum,* though with darker antennal base. Also, the elytra are more parallel-sided, with finer and less punctate striae, and the microsculpture is regular, as in *varium.*

Distribution. Denmark: distributed and rather common, known from all districts. — Sweden: recorded from all districts except P. Lpm. Generally distributed and common except in the northwest. — Norway: rather common in the southern districts, but mainly restricted to coastal areas. Moreover a few fecords from ST. Not found in SF, MR, N, TR and F. — Finland: common all over the country, only lacking in the subalpine-alpine zone. — USSR: north to the Kola peninsula (middle part). — N. and C.Europe, Siberia, the Caucasus.

Biology. A riparian species which usually occurs on the shores of running or stagnant, mainly oligotrophic, fresh waters; rarer along brackish waters. It occurs on very different kinds of soil: clay, mud, peat, etc., usually on soft moist ground with a moderate vegetation of *Carex,* grasses, *Equisetum,* etc. It is a sun-loving species which runs about on bare spots in warm weather. Reproduction takes place in spring.

Subgenus *Notaphemphanes* Netolitzky, 1920

Notaphemphanes Netolitzky, 1920, Koleopt. Rdsch. 8: 96.
    Type-species: *Carabus ephippium* Marsham, 1802.

Entire upper surface, as in *Emphanes,* without microsculpture, but elytral striae complete to apex. Frontal furrows sharper than in *Nothaphus.* Easily distinguished on the pale elytra. Wings full.

113. *Bembidion ephippium* (Marsham, 1802)

Carabus ephippium Marsham, 1802, Ent. Brit.: 462.

2.5-3.0 mm. Black, forebody with a metallic hue, elytra testaceous with indistinct dark fascia behind middle. Appendages pale.

Distribution. In our area only recorded from Denmark: WJ, Skallingen, 1 specimen. On the west coast of Schleswig (S. of the Danish border) some years very abundant, other years missing. — Along the coasts of western Europe (France, Belgium, Holland, BRD, S.England); also S. and SE.Europe, N.Africa.

Biology. A halobiontic species, almost confined to salt marshes near the sea, living at the border of small ponds, etc.; rarely on inland saline localities in C. and SE.Europe (e.g. Neusiedler See).

Subgenus *Emphanes* Motschulsky, 1850

Emphanes Motschulsky, 1850, Käf. Russl.: 12.
Type-species: *Bembidion normannum* Dejean, 1831.

Very small species, very shiny due to total lack of microsculpture. Frontal furrows (Figs 151-153) sharp, parallel (or almost so) between the eyes, prolonged between clypeus in two species. Pronotum convex and narrow, barely wider than head. Wings full.

114. *Bembidion minimum* (Fabricius, 1792)
    Figs 151, 176.

Carabus minimus Fabricius, 1792, Ent. Syst. 1: 168.
Bembidium pusillum Gyllenhal, 1827, Ins. Suec. 4: 403.

2.3-3.2 mm. Black, sometimes with a bluish hue, elytra with apex and/or preapical spot rufous. Appendages piceous, often with paler tibiae, but antennae entirely dark. Frontal furrows (Fig. 151) almost parallel behind the eyes and not prolonged upon clypeus. Pronotum (Fig. 176) clearly wider than head. Elytral striae with moderately strong punctures.

Distribution. Denmark: recorded from all districts, the localities being situated along the coasts of sea and fjords. No or few finds from the NW and NE coasts of Jutland, and the N and W coasts of Zealand. — Sweden: in the south exclusively at the sea coasts. Generally distributed along the W coast (Sk. — Boh.). On the E coast rather distributed on Öl. and Gtl. Accidental on the E coast of Sk.; in Bl., Sm. and Sdm. only a few localities. — Norway: a rare species, local in Ø, Ak and VE, only single records from VA, AA, ST and NT. — Finland: found along the coast in S.Finland, also recorded in Oa; not common. — In the USSR a few records from the southern Karelian Isthmus. — From W.Europe to Siberia.

Biology. A halophilic or halobiontic species, predominantly occurring along the coast in salt marshes, on moist clayey soil with patchy vegetation of e.g. *Salicornia, Atriplex, Obione;* often under seaweed. Less often on inland saline localities; only sporadic appearance at fresh waters. The species is regularly found in company with *Dyschirius salinus* and *Heterocerus flexuosus* Steph., on the North Sea coast also with *Bembidion normannum* and *Pogonus chalceus.* It is a diurnal sun-loving beetle, mainly occurring in spring.

115. *Bembidion normannum* Dejean, 1831
   Fig. 177

Bembidium normannum Dejean, 1831, Spec. Gén. Col. 5: 164.

2.5-3.2 mm. Very similar to *minimum* but more convex, with narrower, somewhat less cordiform pronotum (Fig. 177), the basal margin of which is more elevated. 1st antennal segment and legs paler; elytra diffusely rufous towards apex. Frontal furrows the same. Elytral striae more coarsely punctate.

Distribution. In our area only found in Denmark, where it has a southern distribution. A number of localities are known from the tidal coast from the German border N to Skallingen (WJ). Single localities in EJ (Stavns fjord on Samsø), F (Ristinge, Thurø), LFM (Keldskov), SZ (Korsør Nor), NEZ (Amager). The latter two records are from before 1900. — Distributed with several subspecies along the coasts of Europe to S.Italy.

Biology. Halobiontic, almost confined to coastal salt marshes, occurring on moist usually clayey or silty soil with patchy vegetation of halophytes. Often at the border of small ponds and tidal ditches, also under seaweed. It is frequently found in company with *Dyschirius salinus* and *Bembidion minimum.* Mainly in spring.

116. *Bembidion azurescens* (Dalla Torre, 1877)
   Figs 152, 178, 201.

Bembicidium azurescens Dalla Torre, 1877, 8. Jb. Ver. Naturk. Linz: 54.
Bembidion tenellum auctt.; *nec* Erichson, 1837.

2.5-2.8 mm. In this and the following species the frontal furrows are prolonged upon clypeus (Fig. 152). Black, upper surface with evident, bluish or greenish lustre. Basal antennal segments (at least underneath) and legs rufo-piceous with femora usually darkened. Elytra with apex and as a rule a preapical spot pale. Frontal furrows converging already between the eyes (Fig. 152). Base of pronotum narrower than in *minimum* (Fig. 178). Elytral striae more obsolete towards apex. Penis, Fig. 201.

Distribution. Not in Denmark or Norway. — Sweden: a single specimen known, probably accidental (Gtl., Stånga, 1979). Earlier incorrectly recorded from Halland.

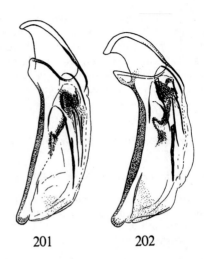

Figs 201, 202. Penis of 201: *Bembidion azurescens* (D. Torre) and 202: *B. tenellum* Er.

201   202

— Not in Finland, but a few records from Vib and southern Kr in the USSR. — From C.Europe to Siberia.

Biology. On river banks on fine sand. In C.Europe predominantly in mountains.

117. *Bembidion tenellum* Erichson, 1837
Figs 153, 181, 202.

Bembidium tenellum Erichson, 1837, Käf. Mark Brandenburg 1(1): 136.
Leja maeotica Kolenati, 1845, Melet. Ent. 1: 79.

2.3-2.8 mm. The only essential characters separating this species from the preceding are: Frontal furrows parallel between the eyes (Fig. 153); base of pronotum oblique laterally (Fig. 181); the structure of the internal sac of penis (Fig. 202). Upper surface with less pronounced metallic hue. The preapical spot less defined or even wanting.

Distribution. Denmark: very rare and usually singly. Nearly all localities are situated in the southern central part of the country. — Sweden: a few specimens were found on the seashore in SE. Skåne (Sandhammaren, 1972, 1977). They were apparently airborn, and the species is not established in Sweden. — Not in Norway. — Not in E.Fennoscandia. — Along the coast of W. and S.Europe; also inland finds in C.Europe.

Biology. Halophilic, living in coastal salt marshes; also found in wind drift on the seashore. In C.Europe frequently in inland saline places.

Subgenus *Leja* Dejean, 1821

Leja Dejean, 1821, Cat. Coll. Col. B. Dejean: 17.
Type-species: *Carabus Sturmii* Panzer, 1805.
Trepanes Motschulsky, 1864, Bull. Soc. Nat. Moscou 37(2): 186.
Type-species: *Carabus articulatus* Panzer, 1795.

Frontal furrows straight, strongly converging forwards, prolonged upon clypeus. Pronotum with 4 small impressions along base (Fig. 182). Elytra with variegated pattern. Wings full.

118. *Bembidion articulatum* (Panzer, 1796)
Fig. 182; pl. 5: 18.

Carabus articulatus Panzer, 1796, Fauna Ins. Germ. 30: 64.

2.9-3.9 mm. A slender species with pronotum not wider than head; its sides parallel posteriorly (Fig. 182). Black, forebody metallic green, very shiny. Elytra with numerous pale spots, confluent at base and near apex. Antennae with at least 3 pale basal segments; legs dark yellow. Elytral striae with much stronger punctures than in *quadrimaculatum*, which is superficially similar.

Distribution. Denmark: recorded from all districts but very sparsely occurring or absent from many areas of western and northern Jutland; in eastern Jutland and on the islands well distributed and rather common. — Sweden: rather distributed and not rare in the south, north to southern Gstr. and southern Dlr.; especially well established in Sk., Öl., Gtl. and in the lowlands around the great lakes in central Sweden. Almost completely lacking at higher altitudes in Sm. and Vg. — Norway: local in the southeast: Ø, AK, HE, VE, TE and AA. — Finland: locally common in the south, recorded north to Ta: Jämsä and Kb: Kitee. — USSR: Vib and southern Kr. — Transpalaearctic, east to Japan.

Biology. On moist clay or clay-mixed soil near fresh water, mainly occurring at the margins of temporary ponds and in clay pits, on barren or sparsely vegetated spots, often hidden in cracks. It is frequently found in company with *B. genei*. Mainly in spring.

119. *Bembidion octomaculatum* (Goeze, 1777)

Carabus octomaculatus Goeze, 1777, Ent. Beytr. 1: 664.
Carabus Sturmii Panzer, 1805, Fauna Ins. Germ. 89: 9.

2.5-2.8 mm. Smaller and stouter than *articulatum*. Black, forebody with a faint metallic hue; elytral spots not confluent at base. Only 1st antennal segment pale; less rufotestaceous. Pronotum shorter and broader, wider than head, sides sinuate posteriorly. Elytra less narrowing apically.

Distribution. Extremely rare and probably not established in Denmark or Sweden. Migrating specimens have been collected at the following localities in Denmark: Bornholm (1857 & 1951) and Sweden: Sk., Sandhammaren (1860, 1969, 1972); Sk., Haväng (1971); Sk., Vitemölla (1972); Sk., Maglehem, Juleboda (1973) and Gtl., Fårön, Sudersand (1946). A single specimen was also collected in Sk., Trelleborg under unknown circumstances more than a 100 years ago. The only specimen not collected close to the sea was found at a small pond (about 1.5 km from the sea) on Gtl., Östergarn (1971). — Norway: no records. — Finland: found twice in sea drift in N: Ekenäs, Klovaskär (1939); once at a small river in N: Tusby (1939) c. 25 km. from the sea. — No records from the Soviet part of E.Fennoscandia. — From Europe and North Africa to W.Siberia.

Biology. In our area only accidentally in wind drift, mainly on the seashore. In C.Europe at the margins of different kinds of fresh water, on silty or sandy soil; rarely on saline localities.

Subgenus *Trepanedoris* Netolitzky, 1918

Trepanedoris Netolitzky, 1918, Koleopt. Rdsch. 7: 24.
Type-species: *Bembidion doris* (Panzer, 1797)

Frontal furrows (Fig. 156) as in subgenus *Trepanes,* that is, absolutely straight, strongly converging, almost meeting at anterior margin of clypeus. Pronotum (Fig. 184) with only one fovea on each side at base. Elytra not variegated. Wings full.

120. *Bembidion doris* (Panzer, 1797)
Figs 156, 184.

Carabus Doris Panzer, 1797, Fauna Ins. Germ. 38: 9.

3.1-3.6 mm. Black, often with a bluish hue, elytra with pale subapical spot (rarely generally rufinistic). At least 1st antennal segment and legs (often except femora) dark rufous. Pronotum (Fig. 184) only slightly wider than head, sides little rounded. External elytral striae obliterating towards apex. Apex of penis with sharp ventral hook.

Distribution. Denmark: very distributed and rather common, recorded from all districts. — Sweden: generally distributed and common in the southern and eastern parts of the country. Absent in the western part from northern Värmland to Torne Lappmark. However, recorded from all districts. — Norway: rather common in the south and south-east, especially in coastal areas; also a few scattered records from HO, ST, NT, N, and TR. Not found in SF, MR and F. — Finland: common and found all over the country except the alpine and subalpine zone. — USSR: found all the way to the Arctic ocean. — From SW.Europe to E.Siberia.

Biology. A very hygrophilous species which occurs mainly at the margins of fresh, both running and stagnant, waters. It is especially abundant along dystrophic and oligotrophic lakes and ponds in rather dense vegetation of *Carex, Eriophorum* and mosses. Also in *Sphagnum* bogs, in company with e.g. *Agonum gracile,* and in shady forest swamps under dead leaves. Often in wind drift on the seashore. Mainly in spring.

Subgenus *Semicampa* Netolitzky, 1910

Semicampa Netolitzky, 1910, Wien. Ent. Ztg. 29: 217.
Type-species: *Bembidion schuppelii* Dejean, 1831.

Frontal furrows (Fig. 154) doubled in anterior part, prolonged upon clypeus. Elytra without pale spots, their striae more or less obliterated apically. Base of pronotum almost straight.

121. *Bembidion schuppelii* Dejean, 1831
Fig. 154.

Bembidium schuppelii Dejean, 1831, Spec. Gén. Col. 5: 860.
Bembidium Sahlbergi Zetterstedt, 1838, Ins. Lapp. 1: 27.

2.8-3.2 mm. Black, upper surface with a blue or green reflection. Only 1st antennal segment entirely pale, legs rufo-testaceous with femora infuscated. Microsculpture evident, reticulate on pronotum, transverse on elytra. Wings usually rudimentary.

Distribution. Denmark: restricted to EJ, and here recorded from 5 localities from Vejle north to Århus. Repeatedly found in numbers on its localities. — Sweden: pronouncedly northeastern, distributed from Nb. to Med., also one locality in Ly. Lpm. Isolated occurrence at the river Klarälven (Vrm.). Not rare in Nb. and at the lower section of the river Klarälven; elsewhere very sparse. — Norway: scattered but locally rather common in all provinces from ST and northwards, except the north-eastern parts of F. Single records from HE and O. — Finland: in the south a few finds on seashores in Ab and N; from Kb and northwards generally distributed but only locally common. — USSR: scattered finds north to southern Kola peninsula. — Europe and Siberia.

Biology. Predominantly on river banks, living on moist silty ground, in rather dense vegetation of e.g. grasses and *Carex,* or in shady sites under bushes. Often seen running on sparsely vegetated spots when it is sunny weather. Rarely occurring on lake- and seashores. In C.Europe often in marshy woodland. It is most abundant in spring, usually hibernating as an imago, but sometimes also as a larva.

122. *Bembidion chaudoirii* Chaudoir, 1850

Bembidion chaudoirii Chaudoir, 1850, Bull. Soc. Nat. Moscou 23(2): 179.

4.0-4.2 mm. Largest species of the subgenus. Piceous black, legs and base of antennae dark rufous; posterior part of elytra with a large but ill-defined dark rufous spot. Pronotum with sides strongly sinuate before hind-angles, with strong keel outside fovea. Microsculpture of head isodiametric, strongly granulate on pronotum with meshes transverse on disc; on the elytra consisting of very fine and dense, non-fusing lines, giving rise to slight iridescence. Wings probably always full. Penis with slender apex without hook. In the inner sclerites quite different from *schuppelii*.

Distribution. Not in Denmark, Sweden, Norway or Finland. — South European USSR and an isolated find on Tshuja Island in the White Sea (several specimens in Mus. Helsinki and Mus. Åbo).

Biology. Probably taken in drift material.

123. *Bembidion gilvipes* Sturm, 1825

Bembidium gilvipes Sturm, 1825, Deutschl. Fauna Ins. 5(6): 149.

2.5-3.0 mm. Black or piceous without any metallic hue, elytra usually paler along suture; 1st antennal segment and legs entirely pale. Upper surface strongly shiny due to lack of microsculpture (except at apex of elytra). Pronotum with narrow base. Elytral striae very coarsely punctate anteriorly but more evenescent posteriorly than in *schuppelii*. Wings either full or completely reduced.

Distribution. Denmark: in Jutland only recent records from SJ; in the eastern districts (F-B) rather distributed, but very local. — Sweden: generally distributed and rather common in the south, north to southern Värmland, southern Dalarna and Gästrikland. Further north rare and recorded only from a few localities in Hls., Med., Jmt. and Ång. — Norway: restricted to south-eastern districts and apparently rare; a few local records from Ø, AK, HE, O and B. — Finland: common in the south, found north to Om. — USSR: common in Vib and southern Kr. — N. and C. Europe, W.Siberia.

Biology. On moist clayey soil, predominantly in marshes and on the shores of fresh waters, usually occurring in somewhat shady sites among litter, e.g. under *Salix* and *Alnus* or in *Phragmites*-vegetation. Also in deciduous forrest swamps with vigorous growth of *Solanum, Lysimachia*, etc. It is sometimes found in abundance in drift material on lake- and seashores. Mainly in spring.

Subgenus *Diplocampa* Bedel, 1896

Diplocampa Bedel, 1896, Cat. Rais. Col. N. Afr. 1: 70.
Type-species: *Bembidium assimile* Gyllenhal, 1810.

At once recognized on the frontal furrows (Fig. 155), which are doubled for their entire length. Elytra almost contantly with one ore more pale spots.

124. *Bembidion fumigatum* (Duftschmid, 1812)
Pl. 5: 10.

Elaphrus fumigatus Duftschmid, 1812, Fauna Austriae 2: 204.

3.5-4.0 mm. Piceous black, forebody with a greenish, elytra with a more bluish, hue. Elytra with numerous distinct pale spots also in anterior half, behind middle usually forming a bent transverse band. Legs and 1 or 2 antennal segments ferrugineous. Pronotum dull from dense, reticulate microsculpture, as in *assimile*. Constantly long-winged.

Distribution. In Denmark recorded from all districts, and seemingly increasing its range and abundance during recent decades. — Sweden: breeding populations occur only in SW. Skåne; some years the species may be rather common here. Accidentally found in SE. Skåne (migrating specimens, sometimes numerous) and on one locality in Halland (Harplinge). — Not in Norway or in East Fennoscandia. — On seashores from W. Europe and the Mediterranean to Transcaspia.

Biology. A halophilic species which mainly occurs in salt marshes near the sea, in rich vegetation on clayey soil, amongst wet debris. Less often along fresh waters in meadows and marshes inland. Sometimes in wind drift on the seashore. Predominantly in spring.

125. *Bembidion assimile* Gyllenhal, 1810
Figs 155, 203.

Bembidium assimile Gyllenhal, 1810, Ins. Suec. 2: 26.

2.8-3.5 mm. Smaller and more convex than *fumigatum*, with a less distinct colour pattern. Black with a greenish hue; elytra as a rule only with the preapical spot and extreme apex pale, in anterior half without or with indistinct spots. Appendages coloured as in preceding species. Microsculpture the same. Antennae and legs stouter. Elytral striae deeper and more coarsely punctate. Wings often reduced.

Distribution. Denmark: known from all districts and rather common; absent from parts of central Jutland. — Sweden: in the southern half, north to Hls., but distribution somewhat scattered. The species is missing in large areas (mainly at higher altitudes) but is generally distributed and common along the west coast as well as on Öl. and Gtl. — Norway: only in southern coastal areas, with scattered records from Ø, AK, VE, TE, VA and R. — Finland: several finds in Al; in N in the Tvärminne area. — USSR: not found in the area covered. — Europe, North Africa, the Caucasus and W. Siberia.

Biology. Predominantly at the margins of eutrophic lakes, ponds and slowly running rivers, on moist clayey or silty soil with luxuriant vegetation of *Carex, Phragmites*, etc., often near the sea. Sometimes under seaweed on beaches. Breeding occurs in spring.

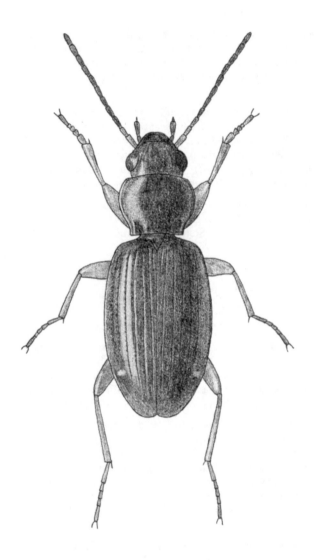

Fig. 203. *Bembidion assimile* Gyll., length 2.8-3.5 mm. (After Victor Hansen).

166

126. *Bembidion clarkii* (Dawson, 1849)
   Fig. 204.

Lopha Clarkii Dawson, 1849, Ann. Mag. Nat. Hist. (2nd Ser.) 3: 215.

3.2-3.7 mm. Best distinguished from the two preceding species on the pronotum, which is broader, less constricted at base, and due to obsolete microsculpture on disc, as shiny as the elytra. Only 1st antennal segment entirely pale. Preapical elytral spot more diffuse, sometimes seemingly absent; only rarely with small pale spots in anterior half. Microsculpture of elytra irregularly transverse, without evident meshes even at apex. Wings almost constantly reduced. Penis (Fig. 204) with short, blunt apex.

Distribution. Denmark: very scattered in Jutland (SJ: one loc., EJ: 4 locs and NWJ: one loc.); well distributed in the eastern districts (F - B). — Sweden: with a limited southern distribution and usually rare. A number of localities are known from S. Skåne, Öland and Gotland. One or a few localities are reported from each of the districts Blekinge, Halland and Småland (eastern part). — Not in Norway or East Fennoscandia. — Western Europe.

Biology. On moist and half-shaded ground in wooded areas, e. g. under *Alnus* and *Salix*, usually occurring at the border of ponds and temporary pools, on clayey mull soil with exuberant vegetation of grasses, *Carex*, etc. Regularly found together with *Acupalpus consputus*, *Agonum livens* and *Badister*-species. Predominantly in May.

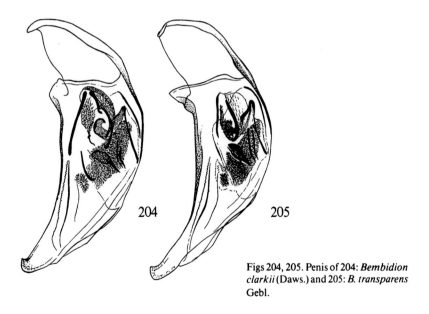

204   205

Figs 204, 205. Penis of 204: *Bembidion clarkii* (Daws.) and 205: *B. transparens* Gebl.

167

127. *Bembidion transparens* (Gebler, 1829)
   Fig. 205.

Peryphus transparens Gebler, 1829, *in* Ledebour: Reise Altai 1: 61.
Bembidion contaminatum J. Sahlberg, 1875, Notis. Sällsk. Faun. Fl. Fenn. Förh. 14: 83.

3.1-3.8 mm. Closely allied to *clarkii*, and earlier regarded as only subspecifically distinct. Also 1st antennal segment usually infuscated. Preapical spot of elytra usually clearly defined. Pronotum somewhat narrower, with less distinct punctuation at base. Elytra with shoulders more prominent and sides more parallel-sided; striae more abbreviated and less evidently punctate. Microsculpture of elytra stronger, notably in the ♀, near apex forming evident, transverse meshes. Wings either full or shorter and narrower than one elytron. Penis (Fig. 205) with long, slender apex and additional differences in the sclerites of the internal sac.

Distribution. Denmark: probably no breeding populations are established in the country. Several times recorded from Bornholm, the beetles occurring under seaweed on beaches, and once also at Bøtø (LFM). — Sweden: in the extreme south (Sk., Hall., Gtl.) very rare and accidental, probably not established. In the lowland around the great lakes in central Sweden (Vg. - Vrm., Ög. - Gstr.) generally distributed and rather common. Also recorded from Ång. and Nb., representing a continuation of the Finnish distribution. — Norway: very rare, only a few local records from N, TR and F. — Finland: rare but widely distributed; most records are from the south-west, but also many from the coast of ObS; otherwise scattered finds north to Li: Inari. — USSR: widely distributed, many records from Lr. — Circumpolar, in Europe south to northern DDR.

Biology. In southern Scandinavia almost confined to the shores of eutrophic lakes, mainly occurring in rich *Phragmites* vegetation, often together with *Odacantha* and *Agonum thoreyi*; rarely on marshy river banks. In the extreme north also living on seashores, whereas in the south it occurs only with sea drift. Mainly in spring.

Subgenus *Bembidion* s. str. Latreille, 1802

Bembidion Latreille, 1802, Hist. Nat. Crust. Ins. 3: 82.
Lopha Dejean, 1821, Cat. Coll. Col. B. Dejean: 17.
   Type-species: *Cicindela quadrimaculata* Linnaeus, 1761.

Base of pronotum (Fig. 179) with a short but deep incision laterally, hind-angles denticulate. Frontal furrows prolonged upon clypeus, simple, moderately convergent. Elytra with pale spots. Wings full.

128. *Bembidion quadrimaculatum* (Linnaeus, 1761)
   Fig. 179; pl. 5: 15.

Cicindela quadrimaculata Linnaeus, 1761, Fauna Suec. ed. 2: 211.
Carabus quadriguttatus Fabricius, 1775, Syst. Ent.: 248; *nec* Olivier, 1795.

2.8-3.5 mm. A small species with very long legs. Black, forebody more or less aeneous, elytra with humeral and almost constantly with preapical yellow spot, and often apex; sometimes also suture bown. Four basal antennal segments and legs rufo-testaceous (or femora slightly infuscated). Palest specimens superficially similar to *articulatum* (but see pronotum Figs. 179, 182). Elytral striae abbreviated, varying in strength of punctuation. Upper surface without microsculpture, very shiny.

Distribution. Denmark: records from all districts except NWJ, but very sparsely distributed in western parts of SJ and WJ and in NEJ. Well distributed and rather common in the eastern part of Jutland and in the island districts. — Sweden: common and widely distributed, but missing in the high mountains of the northeast. Found in all districts except P.Lpm. — Norway: rather common in south-eastern and central districts. Only a few records from eastern TR and F, and apparently lacking in R, HO, SF and N. — Finland: very common all over the country except for the subalpine-alpine zone. — USSR: widely distributed, north to north-western Lr. — Holarctic, with various subspecies in America.

Biology. Rather eurytopic, occurring on sun-exposed, usually only slightly moist, clayey or clay-mixed sandy soil with sparse vegetation. Often in the drier parts of shores and banks of fresh water, in clay pits, etc. Also on cultivated soil, where it may be an important predator on the eggs of certain insect pests. The beetles are active in sunny weather. Often found among drift material on the seashore. Breeding occurs in spring.

129. *Bembidion humerale* Sturm, 1825
Fig. 180.

Bembidium humerale Sturm, 1825, Deutschl. Fauna Ins. 5 (6): 176.

2.6-3.0 mm. Elytra only with a rufo-testaceous spot behind shoulder. Black, forebody with a faint metallic hue. Antennae dark, femora black or piceous, also tarsi darker than the rufo-testaceous tibiae. Antennae shorter and frontal furrows somewhat less convergent than in *quadrimaculatum*. Sides of pronotum more rounded (Fig. 180).

Distribution. Denmark: very limitedly distributed in central parts of Jutland (in SJ, WJ and EJ). Also old records from before 1900 in NEZ; 2 captures on Bornholm (1935 & 1981). — Sweden: distributed in the extreme south (Sk., Bl., Hall., Sm., Gtl., Vg.) but very rare and local. Isolated in Upl. (2 localities). Most of the records are old. — Not in Norway. — Finland: rare in the south, with scattered records from Al, Ab, N, St, Sa, Tb and Om. — USSR: found in Vib and around lake Ladoga (Kr). — From SW. Europe to Romania and NW. Russia.

Biology. Pronouncedly stenotopic, confined to oligotrophic bogs, where it occurs on barren, moderately humid peat soil, often near water. Frequently seen running on the surface in warm sunshine. It is sometimes found in drift material on sea- and lakeshores. Notably in May-June.

130. *Bembidion quadripustulatum* Audinet-Serville, 1821.

Carabus quadriguttatus Olivier, 1795, Ent. 3: 108; *nec* Fabricius, 1775.
Bembidion quadripustulatum Audinet-Serville, 1821, Faune Franc. 1: 80.

3.5-4.0 mm. Darker and larger than *quadrimaculatum*. Black with a bronzy hue, each elytron always with two sharp yellow spots. Antennae black or with base of 3rd and 4th segments pale. Femora black, also tarsi and apex of tibiae infuscated. Elytra broader with more pronounced shoulders and stronger striae.

Distribution. In our area only known from Sweden: accidental captures on seashores on the south coast of Skåne (Skanör in 1970, Gislöv in 1970, Sandhammaren in 1970 and 1977). They certainly represent migrating specimens and the species is not yet established in Sweden. — Central Europe, the Mediterranean area and Turkestan.

Biology. On damp, bare clay or sandy mud.

Subgenus *Philochthus* Stephens, 1828

Philochthus Stephens, 1828, Ill. Brit. Ent. Mand. 2: 2.
Type-species: *Carabus biguttatus* Fabricius, 1779.

At once characterized by the form of the pronotum (Figs 185-187): the sides are rounded to hind-angles, but the base inside these is broadly sinuate. Frontal furrows parallel, not prolonged. Elytra usually with a preapical spot; their microsculpture dense, transverse, causing more or less distinct iridescence.

131. *Bembidion biguttatum* (Fabricius, 1779)
    Fig. 206; pl. 5: 19.

Carabus biguttatus Fabricius, 1779, Reise Norwegen: 232.

3.8-4.3 mm. Distinguished within the subgenus by the 7th elytral stria, which is well developed anteriorly, almost as coarsely punctate as the 6th. Black or piceous, upper surface with strong blue-green reflection, iridescense of elytra very pronounced; preapical macula yellow, apex brown. First antennal segment and legs rufo-testaceous. Pronotum with deep latero-basal sinuation, dull from microreticulation. The latero-basal carina straight. Wings full.

Distribution. Denmark: in Jutland along the eastern coast (SJ + EJ) north to Randers; well distributed in the eastern districts (F - B). — Sweden: widespread and

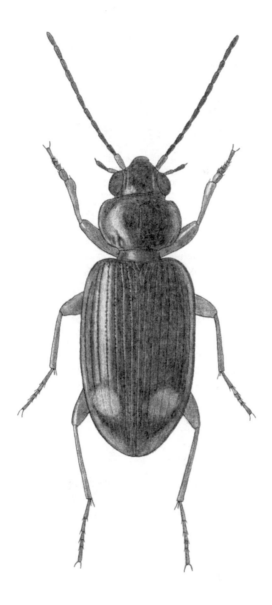

Fig. 206. *Bembidion biguttatum* (F.), length 3.8-4.3 mm. (After Victor Hansen).

common in the western part of Skåne. Furthermore only known from a single locality in Ög. (Örskär, probably accidental). — Not in Norway. — Finland: only found in the southernmost part, north to St: Loimaa and Ta: Ypäjä; sometimes in numbers. — In the USSR a few records from the Karelian Isthmus. — Europe, the Caucasus and W. Siberia.

Biology. On moist mull soil in somewhat shady sites, for instance in eutrophic fens bordering lakes and rivers, usually among tall vegetation. Also in open woodland among moss and leaves at the margins of ponds and pools. Mainly in early spring.

132. *Bembidion iricolor* Bedel, 1879.

Bembidion iricolor Bedel, 1879, Faune Col. Bass. Seine 1: 35.

4.1-5.5 mm. Largest species of the subgenus, closely related to *lunulatum*, and with the same form and microsculpture of pronotum. Antennae slenderer, segments 8-10 more than twice as long as wide. Elytra more stretched, striae more finely punctate in basal half (as in *biguttatum*) but less obliterated towards apex. Coloration as in *lunulatum* except that the ground-colour of elytra usually is more brownish, and the preapical spot indistinct. Wings full.

Distribution. In our area only found in Denmark along the south-western tidal coast from the German border to the Esbjerg-area. — A West European and Mediterranean species; also N. Africa.

Biology. Halobiontic, in our area confined to salt marshes on the North Sea coast of southern Jutland. On clayey or silty soil near the seashore and in inner estuaries; often at the border of drainage canals. Mainly in early spring.

133. *Bembidion lunulatum* (Fourcroy, 1785)

Buprestis lunulatus Fourcroy, 1785, Ent. Paris 1: 51.
Carabus riparius Olivier, 1795, Ent. 3: 115.

3.6-4.1 mm. Shorter and more convex than *biguttatum*. Upper surface only with a faint bluish hue. 1st antennal segment not clearly paler. Preapical spot not always distinct. Head broader, with protruding eyes. Pronotum shiny, disc without microsculpture; latero-basal sinuation shallower and keel outside basal fovea slightly bent outward. Elytral striae with coarser punctures, the 7th even anteriorly rudimentary or wanting. Wings full.

Distribution. Denmark: sparsely distributed and absent from large areas of western and northern Jutland. Has extended its range in recent years. — Sweden: in SW. Skåne found on several localities and has probably increased its abundance during the last 30 years. Also found, but very rarely, in a few localities in SE. Skåne, Blekinge and Öland. — Not in Norway or East Fennoscandia. — C. and S. Europe, North Africa.

Biology. Predominantly on humid clayey soil with rich vegetation, often at the margins of lakes and ponds, in clay pits, etc., also on the sea coast. Rarely on sandy shores and on silty mull soil. Regularly occurring in wind drift on the seashore. It is a late immigrant in Scandinavia and is still expanding. Mainly in spring.

134. *Bembidion aeneum* Germar, 1824
Fig. 185.

Bembidion aeneum Germar, 1824, Ins. Spec. Nov. 1: 28.

3.4-4.5 mm. Black or piceous, upper surface bronze, sometimes bluish. Antennal base indistinctly pale, legs reddish brown; elytra often with apex and sides piceous brown but preapical spot usually indistinct to virtually disappeared (except in transparent light). Pronotum (Fig. 185) broader and flatter than in the three preceding species, entirely micro-reticulate; latero-basal keel straight. Elytral striae finer than in preceding species, faintly punctate, 7th stria barely suggested. The species shows wing dimorphism; though also the rudiment has a reflexed apex, it is shorter and narrower than one elytron.

Distribution. Denmark: scattered records from all districts except Bornholm, but exclusively along the coasts. — Sweden: generally distributed and sometimes common in the western part of S. Sweden, north to Vrm., Dlr. and Hls. In the rest of the country rare or completely lacking, but widely distributed on Öland. — Norway: two separate distribution areas: in the south-east restricted to Ø, AK and VE; moreover in ST, NT and N, here mainly in coastal areas. — Finland: not recorded. — USSR: recorded in 2 specimens at Paanajärvi at the Oulanka river in northernmost Kr. — Disjunctly distributed, one area in W. Europe, another from Romania to the Caspian Sea.

Biology. A halophilic or halobiontic species, occurring on firm, moist clay soil with often rich but short vegetation of grasses, *Carices*, etc. It is primarily a seashore inhabitant, living above the high water mark, often far from the water; also along river estuaries. In southern Sweden also in the inland, e.g. on clayey lake shores and river banks, even in agricultural fields. The beetles often hide in clay cracks, or run about on the surface in sunshine. Predominantly an imaginal hibernator, but also larvae sometimes hibernate.

135. *Bembidion guttula* (Fabricius, 1792)
Figs. 186, 207.

Carabus guttula Fabricius, 1792, Ent. Syst. 1: 166.

2.8-3.5 mm. This and the following species are the smallest of the subgenus and the base of pronotum is only faintly sinuate laterally (Fig. 186). The two species are easily confused, mainly due to a considerable individual variability. *B. guttula* is black or piceous, elytra iridescent and with a more or less pronounced bluish or greenish ground

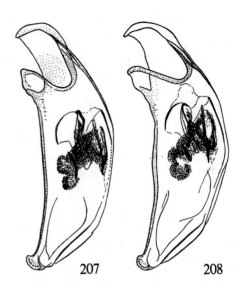

Figs 207, 208. Penis of 207: *Bembidion guttula* (F.) and 208: *B. mannerheimii* Sahlbg.

207

208

lustre; preapical spot from sharp to very diffuse, also apex normally pale. Base of antennae more or less pale (but sometimes even 1st segment infuscated); legs dark testaceous or rufous, usually with infuscated femora. Pronotum (Fig. 186) entirely microreticulate, its sides almost straight behind middle. Sides of elytra little rounded. Wings either full or reduced, in the latter case at least reaching base of 5th abdominal tergite. Penis, Fig. 207.

Distribution. Denmark: rather distributed and common, known from all districts. — Sweden: widespread; common in the southern half of the country, in the northern half less common and apparently absent from the western mountains (not recorded from Hrj. and entire Lapland). — Norway: a south-eastern species restricted to Ø, AK, HE, O, B, VE, TE, VA and AA, but locally rather common. — Finland: found in all provinces except Ks, LkW, Le and Li; very common in the south. — USSR: found north to Murmansk. — Europe, North Africa, the Caucasus and Siberia; introduced in North America.

Biology. On moist clay or clay-mixed soil near fresh water, usually in open land in rich vegetation of *Carices*, grasses, etc., but also in light deciduous forest, e.g. at the margins of ponds and temporary pools. Mainly in spring.

136. *Bembidion mannerheimii* Sahlberg, 1827
    Figs. 187, 208.

Bembidium Mannerheimii Sahlberg, 1827, Ins. Fenn. 1: 201; *nec* Dejean, 1829.

174

Bembidion unicolor Chaudoir, 1850, Bull. Soc. Nat. Moscou 23 (2): 176.
Bembidion haemorrhoum auctt.; *nec* Stephens, 1828.

2.8-3.4 mm. Broader and more convex than *guttula*, pronotum (Fig. 187) and elytra with sides more rounded. Upper surface pure black or elytra faintly iridescent. 1st antennal segment and legs more clearly rufo-testaceous, femora not darker. Elytra, in normal light, without or with very diffuse preapical spot. Antennae somewhat stouter. Wing rudiment not surpassing 2nd abdominal tergite; full-winged specimens extremely rare. Penis, Fig. 208.

Distribution. Denmark: widely distributed and common, known from all districts. — Sweden: generally distributed from the south to 60°N and rather common; also scattered localities between 60°-63°N but here restricted to the eastern part of the country. — Norway: a southern species, mainly restricted to coastal areas from AK to HO. — Finland: rather common in the south, found north to Ok. — USSR: found north to the White Sea. — Europe and Siberia.

Biology In moderately moist shady habitats. Found among litter and moss on sparsely vegetated spots, notably in deciduous woodland on mull soil, e.g. in the drier parts of alder swamps and at the margins of brooks and ponds. Also in rich fens under bushes of *Salix*, etc.; rarely in oligotrophic bogs with *Sphagnum*. Mainly in spring.

Subgenus *Cillenus* Samouelle, 1819

Cillenus Samouelle, 1819, Ent. Comp.: 148.
   Type-species: *Cillenus lateralis* Samouelle, 1819.

Often regarded as separate genus and at once recognized on the 4 dorsal punctures of elytra. Head very broad without neck (Fig. 209). Mandibles protruding. Antennae short, moniliform (with rounded outer segments). Entire upper surface with strong reticulate microsculpture.

137. ***Bembidion laterale*** (Samouelle, 1819)
   Fig. 209; pl. 5: 17.

Cillenus lateralis Samouelle, 1819, Ent. Comp.: 148.

3.0-4.0 mm. Upper surface testaceous, except that the head is often dark piceous. Forebody, notably the head, with a greenish metallic reflection. Elytra usually with a longitudinal, faintly dark spot. Appendages pale. Wings strongly reduced (acc. to Ganglbauer, macropterous specimens are known).

Distribution. In our area only recorded from Denmark: several localities are known on the tidal south-western coast of Jutland from Ribe to Skallingen (SJ + WJ). Also a record from EJ, Vorsø in Horsens Fjord, in great numbers 1980/81; see below. — On the coasts of W. Europe.

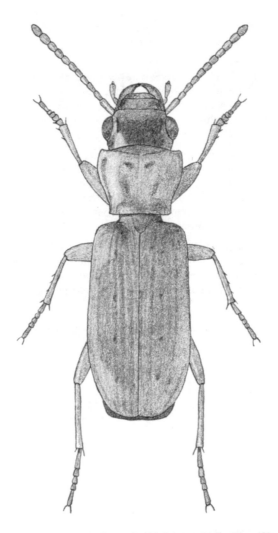

Fig. 209. *Bembidion laterale* (Sam.), length 3-4 mm. (After Victor **Hansen**).

Biology. A very stenotopic, halobiontic species, which is confined to tidal salt marshes, occurring on wet, well-drained, sparsely vegetated or barren sands or muddy sands, regularly flooded by the tide. During inactivity, e.g. at high tide, the beetles stay in burrows in the sand or hide under stones etc.; they are often active on the surface in warm sunshine. Adults *B. laterale* mainly feed on staphylinid beetles and amphipods,

176

e.g. *Bledius fergussoni* Joy (*arenarius* Payk.), *Diglotta*-species, *Corophium volutator* Pallas and *Talitrus saltator* Montagu. Breeding occurs in spring.

Note (by P. Jørum). A variant of *B. laterale* occurs on the isle Vorsø in Horsens Fjord (EJ, Denmark). It differs from normally coloured speciemens in the following characters: forebody, including pronotum, and the longitudinal spots on elytra are dark piceous, with a more or less pronounced greenish metallic lustre. Under surface entirely dark.

Subgenus *Nepha* Motschulsky, 1864

Nepha Motschulsky, 1864, Bull. Soc. Nat. Moscou 37(2): 190.
Type-species: *Bembidion menetriesi* Kolenati, 1845.

A strongly shiny species with four-spotted elytra and narrow pronotum not wider than long (Fig. 183). Elytral striae disappearing behind middle. Frontal furrows parallel.

138. *Bembidion genei* Küster, 1847
Fig. 183; pl. 5: 14.

Bembidium Genéi Küster, 1847, Käf. Eur. 9: 21.
Elaphrus quadriguttatus Illiger, 1798, Verz. Käf. Preuss.: 233; *nec* (Fabricius, 1775).
Bembidion genei illigeri Netolitzky, 1914, Ent. Blätt. 10: 54.

4.0-4.9 mm. Black, forebody usually with a greenish hue; each with a sharp humeral and preapical spot. Base of antennae (at least underneath) and legs (except tip of femora and base of tibiae) testaceous. Pronotum scarcely wider than head, base (Fig. 183) without or with a barely suggested latero-basal keel. Upper surface without microsculpture. Wings full.

The nominate *genei* Küster has a Mediterranean distribution and our form is ssp. *illigeri* Netolitzky, 1914.

Distribution. Denmark: well distributed and rather common, however sparse in western Jutland; records from all districts. — Sweden: from Sk. to Dls., Vstm. and Upl. but absent from the southern highland from N. Skåne to S. Västergötland and Östergötland. Locally common. — Not in Norway. — Finland: several localities in Al; once recorded in sea drift in N: Tvärminne (1939). — USSR: no records from Vib, Kr or Lr. — Europe and the Mediterranean area (several subspecies).

Biology. On open moist clay or clay-mixed sandy soil lacking or with sparse vegetation, e.g. *Tussilago*, usually near water; often on spots where the upper surface is somewhat dried-up and cracked. It occurs mainly in clay pits and on clayey banks of ponds, lakes, rivers, etc., also on slopes with seeping water, on the sea coast as well as inland, often in company with *B. articulatum* and *nitidulum*. Breeding takes place in spring.

210 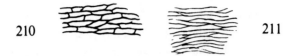 211

Figs 210, 211. Microsculpture near apex of elytra of *Bembidion*. — 210: *fellmanni* Mann. and 211: *difficile* Mtsch.

Subgenus *Plataphodes* Ganglbauer, 1892

Plataphodes Ganglbauer, 1892, Käf. Mitteleur. 1: 160.
Type-species: *Bembidion Fellmanni* Mannerheim, 1823.

The internal structures of the penis demonstrate the close relationship with the following subgenus *(Plataphus)*. They have also the sharp, complete, or almost complete, elytral striae in common. The only reliable external difference is found in the development of the shoulder, which, in *Plataphus*, is completely rounded (Fig. 190), whereas in *Plataphodes* (Fig. 189) the marginal elytral bead is connected with an arcuate rudiment of the basal bead. Small, entirely dark species. Wings full.

139. Bembidion fellmanni Mannerheim, 1823
 Figs 210, 212.

Bembidion Fellmanni Mannerheim, 1823, *in* Hummel: Essais Ent. 3: 43.
Bembidion palmeni J. Sahlberg, 1900, Acta Soc. Faun. Fl. Fenn. 19: 3.

3.7-4.4 mm. A flat species. Black, upper surface almost constantly with a brass, green or blue lustre. Appendages dark. Pronotum strongly widening forwards. Elytra broader apically, with rather deep, faintly punctate striae. Microsculpture of elytra (Fig. 210), though sometimes variable, always forms evident transverse meshes at least near apex, especially in the ♀. Penis, Fig. 212.

Distribution. Not in Denmark. — Sweden: northern and restricted to two rather well separated areas in the high mountains (Hls., Hrj. and Jmt. in the south; P.Lpm., Lu.Lpm. and T.Lpm, in the north). Local but not rare. — Norway: scattered, but locally common in the north, in the south mainly restricted to central alpine areas. Single records also from B and R. — Finland: northern, found south to Ks: Kuusamo. — USSR: Lr and northernmost Kr. — From N. Europe to E. Siberia, in C. Europe found only in Rumania.

Biology. In the fjelds, mainly in the alpine region and the birch zone. On sterile gravel of lake-shores or along brooks among gravel and pebbles; often associated with *B. hastii*. A quite different habitat is on alpine meadows.

140. **Bembidion difficile** (Motschulsky, 1844)
   Figs 211, 213.

Peryphus difficilis Motschulsky, 1844, Ins. Sibér.: 248.
Bembidion aeruginosum auctt.; *nec* Gebler, 1833.

3.8-4.5 mm. Difficult to separate from *fellmanni* without investigation of the male genitalia. Somewhat stouter, more convex and with shorter legs. The metallic colour is often more greenish on the sides of pronotum and elytra. Pronotum less narrowed towards the base, which is more rectilinear. The elytral microsculpture (Fig. 211) is always denser and consists of transverse lines without formation of defined meshes. Penis (Fig. 213) stouter; internal sac with a characteristic plate but without the stylet present in *fellmanni*.

Distribution. Not in Denmark. — Sweden: rather distributed and often common in the north, south to northern Värmland. Occurs at lower altitudes than *fellmanni*, but

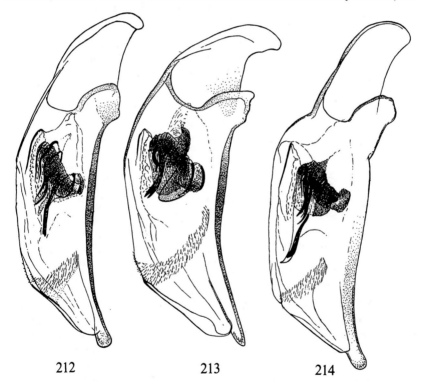

212   213   214

Figs 212-214. Penis of *Bembidion*, subg. *Plataphodes*. — 212: *fellmanni* Mann.; 213: *difficile* (Mtsch.); 214: *crenulatum* F. Sahlbg. (holotype of *ponojense* J. Sahlbg.).

179

not near the coast. — Norway: rather common in the north; very local in central districts (MR, ST). In the south scattered in continental areas (HE, O, B). — Finland: rather common in the north, found south to Ok. — USSR: Lr and northern Kr. — From North Europe to Siberia and Mongolia; in Central Europe found in the Tatra Mts.

Biology. Characteristic for the northern forest region, and usually not found above the timber limit. Mainly along running, often quite small, waters, less often on lake- and seashores. It occurs on moist silty sand, gravel, or clay with sparse vegetation of *Carices*, grasses, moss, etc.; notably in somewhat shady sites such as under *Salix*-bushes. Mainly in summer.

141. ***Bembidion crenulatum*** F. Sahlberg, 1844
    Fig. 214.

Bembidion crenulatum F. Sahlberg, 1844, Nov. Ochotsk Car. spec.: 58.
Bembidion ponojense J. Sahlberg, 1876, Notis. Sällsk. Faun. Fl. Fenn. Förh. 14: 75.

3.7-4.4 mm. Very similar to *fellmanni* and *difficile*. General habitus as in *difficile*, that is, with pronotum broad and little constricted at base. Microsculpture the same, but often with more suggested reticulation on the disc of pronotum, the base of which has a shallow but evident sinuation laterally. The only reliable difference between the two species lies in the inner armature of the penis (Fig. 214).

Distribution. Not in Denmark, Sweden, Norway or Finland. — USSR: Ponoj in eastern Kola Peninsula. — A tundra species, east to the Bering Strait and the Ochotsk Sea, but distribution imperfectly known.

Subgenus *Plataphus* Motschulsky, 1864

Plataphus Motschulsky, 1864, Bull. Soc. Nat. Moscou 37 (2): 184.
    Type-species: *Elaphrus prasinus* Duftschmid, 1812.

Middle-sized species, uniformly dark (or with diffusely rufinistic elytra). Pronotum with sharp latero-basal carina. Elytral striae sharp to apex, 7th stria evident but not reaching apex, shoulders evenly rounded (Fig. 190). Entire upper surface microsculptured. Metasternal process between meso-coxae only margined laterally. Wings full.

Two species (*virens, hastii*) have a fringe of bristles along hindmargin of the abdominal sternites and have therefore sometimes been referred to a separate subgenus, *Blepharoplataphus* Netolitzky (cf. Figs 192, 193).
    All species are strictly riparian.

142. **Bembidion prasinum** (Duftschmid, 1812)
Figs 193, 215.

Elaphrus prasinus Duftschmid, 1812, Fauna Austriae 2: 201.
Bembidium olivaceum Gyllenhal, 1827, Ins. Suec. 4: 408.

4.2-5.5 mm. Best recognized on the virtually impunctate elytral striae. Broad and flat with parallel elytral sides. Black with a faint greenish lustre, bluish underneath; elytra often rufinistic (*"kolstroemi"* Sahlberg). 1st antennal segment, at least underneath, and base of femora rufous. Inner elytral striae with extremely faint rudiments of punctures. Abdominal sternites only with the normal pair of setae (Fig. 193); see footnote on p. 141. Penis (Fig. 215) with very characteristic armature.

Distribution. Not in Denmark. — Sweden: rather distributed in the north south to N.Uppland, S.Dalarna and N.Värmland; usually frequent. — Norway: widely distributed, but local. Not recorded from VE, HO and MR. — Finland: northern, found south to ObN: Kemi. — In the USSR in Lr and northernmost Kr. — Europe and Siberia.

Biology. On sterile gravelly and stony river banks, rarely along brooks and on lake shores, always close to water. It is often found in company with *B. virens* and *saxatile*. Mainly inhabiting the coniferous region, but also occurring in the birch forests, occasionally just above the timber limit.

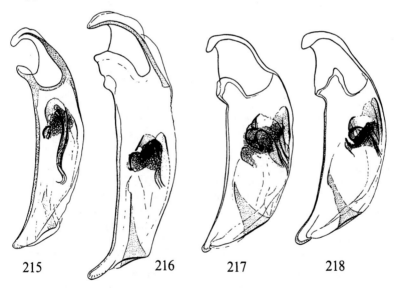

215   216   217   218

Figs 215-218. Penis of *Bembidion,* subg. *Plataphus.* — 215: *prasinum* (Dft.); 216: *hyperboraeorum* Munst.; 217: *virens* Gyll.; 218: *hastii* Sahlbg.

143. *Bembidion hyperboraeorum* Munster, 1923
Fig. 216.

Bembidion hyperboraeorum Munster, 1923, Norsk Ent. Tidsskr. 1: 238.

4.1-5.3 mm. Pure black, including elytra and appendages, usually with a faint aeneous hue. Pronotum less narrowing towards base than in *prasinum*. Elytra shorter, more convex, somewhat widening towards apex, which is more broadly rounded; striae with faint but evident punctures. Penis (Fig. 216) quite different from that of *prasinum*.

Distribution. Not in Denmark. — Sweden: a usually rare northern species, restricted to the western mountains of Jämtland and Lappland. — Norway: a northern species, locally common in TR, F, and the northern parts of N. — Finland: northern, southernmost record from Ok: Sotkamo. — USSR: Lr and northernmost Kr. — Circumpolar.

Biology. In the mountains, from the upper coniferous region to the alpine zone, occurring close to water on sterile gravelly or stony, sometimes sandy banks of rivers, brooks and lakes. Often associated with *B. hastii*. Predominantly in late spring and summer, mostly hibernating as an imago, perhaps also as a larva.

144. *Bembidion virens* Gyllenhal, 1827
Figs. 192, 217.

Bembidion virens Gyllenhal, 1827, Ins. Suec. 4: 407.
Bembidion Pfeiffii Sahlberg, 1827, Ins. Fenn. 1: 195.

4.5-5.4 mm. Abdominal sternites with a fringe of bristles along hind-margin (Fig. 192). Black, upper surface with a pronounced green or brassy lustre; appendages black, only trochanters and sometimes innermost part of femora brown. Pronotum and elytra more convex than in *prasinum*, the former with rounded sides, the latter considerably widening apically. Elytral striae with coarser punctures than in *hyperboraeorum*. Penis, Fig. 217.

Distribution. Not in Denmark. — Sweden: scattered distributed in the north, missing over large areas and only locally common. The southern limit of distribution runs through Hälsingland and northern Värmland, but also isolated localities on the coasts in Bohuslän, Uppland and Gästrikland. — Norway: widely distributed and rather common in all districts except MR. — Finland: northern, found south to Ks: Kuusamo; also isolated finds on the coast of Om and ObS. — Lr and northern Kr of the USSR. — N. Europe and Siberia; also an isolated population at Lake Geneva.

Biology. Mostly on sterile, gravelly or stony river banks and lake shores, rarely on seashores, always near the water edge. In the mountains usually below the timber line. It is often found in company with *B. saxatile*, *prasinum* and *bipunctatum*. Breeding occurs in spring.

145. *Bembidion hastii* Sahlberg, 1827
Figs 170, 218.

Bembidium Hastii Sahlberg, 1827, Ins. Fenn. 1: 195.

4.2-5.4 mm. Slenderer and flatter than *virens*. Black, upper surface with a bluish, rarely greenish, metallic lustre. Femora, at least at base, sometimes entire legs, rufous; 1st antennal segment often brown underneath. Pronotum (Fig. 170) more stretched than in *virens*, with sides almost parallel posteriorly. Elytra less widening apically, striae usually stronger and more coarsely punctate, inner intervals more or less convex. Microsculpture of elytra stronger. Penis, Fig. 218.

Distribution. Not in Denmark. — Sweden: only in the north, generally distributed in the high mountains south to Hrj. and 3 localities at the coast of Nb. Often very frequent. — Norway: common in the north and in southern alpine areas, in ST, NT and N scattered along the Swedish border. Isolated records also from inner parts of R. — Finland: northern, found south to ObN: Rovaniemi and Ks: Kuusamo; also along the coast in Om and ObS. — Lr and northern Kr in the USSR. — Circumpolar.

Biology. Notably on barren, gravelly or stony sites along rivers, brooks and lakes, in northernmost Fennoscandia also on seashores, always close to the water edge. It is most predominant in the alpine and birch regions of the mountains, often occurring together with *B. fellmanni*. Also on the tundra of the Kola peninsula. Breeding takes place in July-September, both larvae and imagines can hibernate (Andersen 1970).

Subgenus *Hirmoplataphus* Netolitzky, 1942

Hirmoplataphus Netolitzky, 1942, Koleopt. Rdsch. 28: 107.
Type-species: *Bembidion hirmocaelum* Chaudoir, 1850.

Related to *Plataphus* but with a thin or rudimentary latero-basal carina of pronotum. Elytral striae complete, strongly punctate anteriorly, the 7th reaching apex. Metasternal process (between meso-coxae) entirely unmargined, even laterally.

146. *Bembidion hirmocaelum* Chaudoir, 1850

Bembidion hirmocaelum Chaudoir, 1850, Bull. Soc. Nat. Moscou 23 (2): 190.

4.5-5.2 mm. Stout species with broad base of pronotum and hindangles protruding, about right. Black, upper surface more or less brassy, 1st antennal segment dark rufous, legs piceous. Microsculpture obsolete on disc of pronotum in the ♀, absent in the ♂; on the elytra consisting of extremely fine transverse lines in both sexes.

Distribution. Not in Denmark, Sweden, Norway or Finland. — USSR: found in the Svir area. — Siberia.

Biology. On barren, sandy or stony banks of rivers and brooks.

Subgenus *Peryphus* Dejean, 1821

Peryphus Dejean, 1821, Cat. Coll. Col. B. Dejean: 17.
 Type-species: *Bembidium tetracolum* Say, 1823.
Peryphiolus Jeannel, 1941, Faune de France 39: 496.
 Type-species: *Bembidium monticola* Sturm, 1825.

This is the largest subgenus of *Bembidion*. It contains middle-sized species, with elytral striae more or less obsolete towards apex (least so in *saxatile*); 7th stria from absent to faint (weaker than 6th). As to coloration the elytra are either uniformly dark or with pale spots, usually (except in *lunatum*) with a subhumeral spot and a preapical (or apical) spot on each elytron. Pronotum pronouncedly cordate with sides sinuate in front of hind-angles. Elytra with evident microsculpture. Antennal base more or less pale. Meta-sternal process (between meso-coxae) with complete border (Figs 227, 228). Wings sometimes reduced. Penis much varying, inner armature reliable for species identification.

The two first species are often referred to subgenus *Bembidionetolitzkya* E. Strand (*Daniela* Net.), distinguished among other things, by the complete 2nd elytral stria.

147. *Bembidion tibiale* (Duftschmid, 1812)
 Fig. 219.

Elaphrus tibialis Duftschmid, 1812, Fauna Austriae 2: 209.

5.5-6.5 mm. Large and flat with long parallel-sided elytra. Upper surface with blue or green reflection. Appendages piceous but 1st antennal segment, tibiae, tarsi and often apex of femora pale. Frontal furrows deep. Pronotum with base impunctate, laterobasal carina sharp. Elytral striae with weaker punctures than in *nitidulum* and *stephensi*. Microsculpture strong, on the elytra forming dense, very transverse meshes. Penis (Fig. 219) large and stout with well developed internal armature.

Distribution. In our area only found in Norway: a few old record records from R. — C. and S. Europe.

Biology. On sterile, stony banks along brooks and rivers, usually in somewhat shaded sites.

Figs 219-225. Penis of *Bembidion*, subg. *Peryphus*. — 219: *tibiale* (Dft.); 220: *mckinleyi scandicum* Lth.; 221: *nitidulum* (Marsh.); 222: *yukonum* Fall; 223: *grapii* Gyll.; 224: *dauricum* (Mtsch.); 225: *stephensi* Crotch.

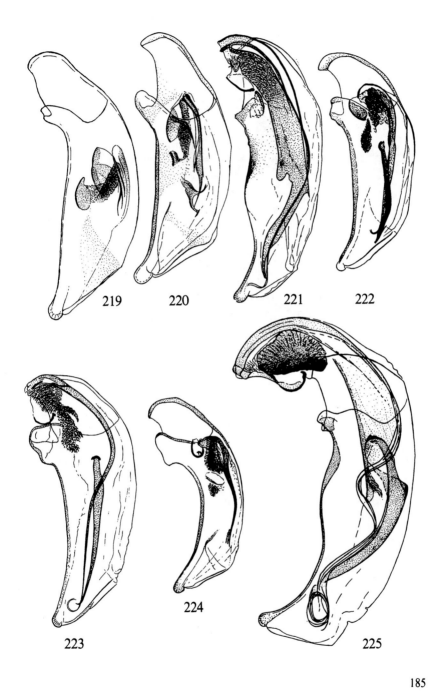

219 220 221 222

223

224

225

185

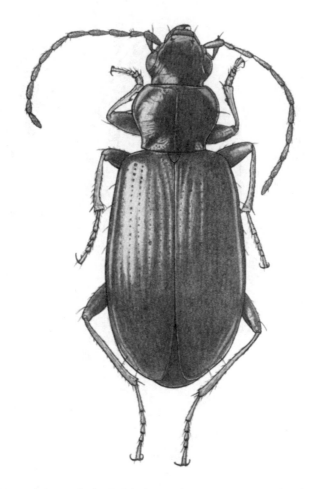

Fig. 226. *Bembidion mckinleyi* Fall, holotype of ssp. *scandicum* Lth., length 4.5-5 mm.

148. *Bembidion mckinleyi* Fall, 1926
   Figs 162, 220, 226.

Bembidion macropterum auctt.; *nec* J. Sahlberg.
Bembidion mckinleyi Fall, 1926, Pan-Pacific Ent. 2: 132.
Bembidion scandicum Lindroth, 1943, Ent. Tidskr. 64: 4.

4.5-5.0 mm. Pronouncedly slender (elytra about 4 times the length of the pronotum)

186

and similar in habitus to *hastii*, but with apically incomplete striae. Piceous black, elytra with a faint, pronotum with a stronger, greenish bronze hue; elytra usually slightly rufinistic anteriorly. At least 1st antennal segment, tibiae, tarsi and extreme apex of femora more or less brown. Pronotum (Fig. 162) strongly widened anteriorly, hind-angles prominent, right or slightly acute, latero-basal carina rudimentary, base rugulosely punctate. Microsculpture evident on entire frons, suggested also on centre of pronotum and clearly reticulate over entire elytra (also on the base of the ♂). Penis (Fig. 220) with characteristic "triconed body".

Our form is subspecies *scandicum* Lindroth, otherwise known only from Siberia. In N. America occur two separate subspecies *mckinleyi* s.str. and *carneum* Lindroth. They are separated, among other things, on the internal sac of penis (Lindroth 1963: Fig. 150).

Distribution. Not in Denmark. — Sweden: extremely restricted in distribution, known only from the Abisko area (T.Lpm.); here sometimes found to be frequent. — Norway: rare, mainly restricted to inner parts of TR; single records only from N and F. — Finland: a few records from Le (the Kilpisjärvi area) since 1977. — USSR: not found in the area covered. — Almost circumpolar; in North America west of the Hudson Bay. Siberia: Baical.

Biology. Along small rivers in the birch and coniferous zones of the fjelds, occurring on gravelly or stony, sometimes sandy, ground, on exposed or somewhat shaded banks. The species has been found in company with, among others, *B. hastii, hyperboraeorum* and *prasinum*. Oviposition occurs from about late June until August; larvae, and perhaps also imagines, hibernate (Andersen 1970b).

149. *Bembidion monticola* Sturm, 1825
Fig. 168.

Bembidium monticulum Sturm, 1825, Deutschl. Fauna Ins. 5 (6): 135.

4.5-5.0 mm. By Jeannel referred to the separate subgenus *Peryphiolus*, due to the somewhat more evident 2nd elytral striae and the strong pronotal microsculpture. Bluish greeen, dull, 1st antennal segment and legs rufo-testaceous (or femora very fainty darkened). Pronotum (Fig. 168) only slightly wider than head. Elytra more stretched than in *nitidulum* and more acuminate; striae more weakly punctate. Microsculpture of pronotum reticulate (as in *bruxellense*). On the elytra densely transverse striae without forming evident meshes.

Distribution. Denmark: very rare; WJ: Varde å in Nørholm skov, 7 specimens 1920-24; EJ: Grejså i Grejsdalen, 1 specimen 1920. — Not in Sweden or Norway. — Finland: found in Ab: Pusula (2 specimens); also reported from Kb: Juuka. — USSR: found in Vib. — Europe, the Caucasus.

Biology. Along small rivers and brooks, usually occurring on clay or fine sand covered with small stones, on sparsely vegetated or barren banks in somewhat shaded situations, e.g. in deciduous forest. In C. Europe and Britain predominantly in mountains.

150. Bembidion nitidulum (Marsham, 1802)
Fig. 221; pl. 5: 9.

Carabus nitidulus Marsham, 1802, Ent. Brit.: 454.
Bembidium rufipes Gyllenhal, 1810, Ins. Suec. ed. 2: 18.

4.5-5.3 mm. This and the four following species have unicolorous dark elytra and are thereby different from all following species of subgenus Peryphus (except decorum). B. nitidulum has constantly a vivid bluish green or almost blue upper surface; ground colour black, underside with weak bluish hue; antennae with 1 or 2 basal segments entirely and the 2 following at base, as well as palpi and legs, rufo-testaceous, but femora and penultimate segment of maxillary palpi infuscated. Pronotum as in bruxellense, upper surface transversely microsculptured. Elytra with striae deep, strongly punctate in basal half; the reticulate microsculpture evident near apex only. Penis, Fig. 221.

Distribution. Denmark: distributed and rather common in eastern parts of Jutland and on the islands. Also isolated records from NWJ (Fur and Bulbjerg) and NEJ (Svinkløv). — Sweden: scattered distributed in the western parts from Sk. to Ly.Lpm., usually rare. Completely absent from eastern Sweden and over large areas of central Sweden. — Norway: in most districts, but scattered and very local. Not in TR and F. — Finland: a few localities in the eastern parts, Ka: Imatra, Sa: Lemi and Kb: Liperi and Polvijärvi. — USSR: many records from the Karelian Isthmus, also from the Svir area. — Europe and North Africa.

Biology. Confined to moist clay or clay-mixed soil, often with trickling water, usually occurring on open ground with sparse vegetation of Tussilago, etc., often in clay and gravel pits, on sea slopes and river banks. Also in moderately shady habitats in deciduous forest. The beetles often hide in cracks. They reproduce in spring.

151. Bembidion grapii Gyllenhal, 1827
Figs 171, 223, 228, 229.

Bembidium grapii Gyllenhal, 1827, Ins. Suec. 4: 403.
Bembidium sahlbergii Dejean, 1831, Spec. Gén. Col. 5: 144; nec Zetterstedt, 1840.

4.0-4.8 mm. Reminding of a large lampros but with shoulders rounded (cf. Fig. 190). Black or dark piceous, upper surface with a faint bronze hue (rarely bluish or greenish). Legs brownish or rufous brown with femora sometimes slightly infuscated. Antennae (Fig. 229) shorter than in nitidulum and yukonum and metathoracic process more pointed (Fig. 228). The microsculpture of elytra transverse also at apex without meshes. Penis, Fig. 223. Hind-wings dimorphic, in Norway mainly brachypterous, in Sweden and Finland predominantly macropterous.

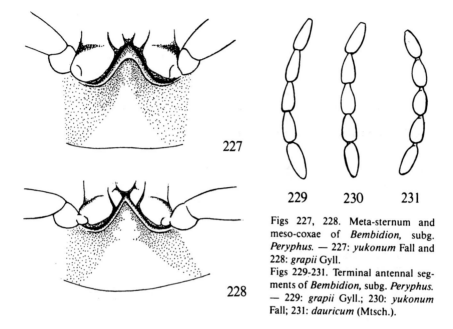

Figs 227, 228. Meta-sternum and meso-coxae of *Bembidion,* subg. *Peryphus.* — 227: *yukonum* Fall and 228: *grapii* Gyll.
Figs 229-231. Terminal antennal segments of *Bembidion,* subg. *Peryphus.* — 229: *grapii* Gyll.; 230: *yukonum* Fall; 231: *dauricum* (Mtsch.).

Distribution. Not in Denmark. — Sweden: generally distributed and not rare in the northern half of the country, south to northern Värmland. Not recorded from coastal regions. — Norway: scattered and local in N, TR and F; further south only isolated records from HE, O, VA, R and ST. — Finland: rare but found over most of the country except the western coastal areas; in the south only a few scattered records, some of them on the coast or the islands. — USSR: scattered finds from Vib to the Arctic coasts. — Circumpolar.

Biology. Confined to the high boreal coniferous and birch regions, living on fine clay-mixed sand, sligthly moist on surface and with sparse vegetation (e.g. *Festuca ovina*), especially on spots with tiny moss. Usually in somewhat shaded sites, e.g. at the edge of forests or on northern slopes. It is often associated with *Miscodera* and *Trichocellus cognatus*. Mainly in summer.

152. *Bembidion yukonum* Fall, 1926
Figs 222, 227, 230.

Bembidion yukonum Fall, 1926, Pan-Pacific Ent. 2: 131.
Bembidion grapeioides Munster, 1930, Norsk Ent. Tidsskr. 2: 354.
Bembidion sahlbergioides Munster, 1932, Norsk Ent. Tidsskr. 3: 82.

4.2-5.6 mm. Closely allied to *grapii* but usually with a more pronounced greenish or aeneous lustre; otherwise coloured the same. Antennae slenderer (Fig. 230). Pronotum a little more constricted at base. Elytral striae, notably 7th, with larger punctures. Metasternal process, Fig. 227. Microsculpture of pronotum more reduced than in *grapii*, weak and restricted to sides even in the ♀, lacking in the ♂; more reduced also on the elytra but with evident formation of meshes (in the ♂ developed at apex only). Wings varying. Penis (Fig. 222) with inner armature much shortened.

Distribution. Not in Denmark. — Sweden: very rare, only a few localities in the extreme northwestern part (T.Lpm.). — Norway: very rare, only a few records from F. — Finland: only a few records from Li: Utsjoki. — USSR: Lr (Pechenga and Ponoj). — Circumpolar.

Biology. Only in the fjelds. Mainly on cracked clayish moving soil on underlying moraine in the birch region. Both on river banks and far from water. Mainly in summer.

153. *Bembidion dauricum* (Motschulsky, 1844)
Figs. 172, 224, 231.

Leja daurica Motschulsky, 1844, Mem. Acad. Sci. Math.-Phys. Nat. St. Petersb. 5: 256.
Bembidion pseudoproperans Netolitzky, 1920, Koleopt. Rdsch. 8: 69.
Bembidion lysholmi Munster, 1930, Norsk Ent. Tidsskr. 2: 353.

3.8-4.3 mm. A small species, easily separated from *grapii* and related species by the reticulate microsculpture. Coloured as *grapii*, antennae as in its darkest form, that is, often also with 1st antennal segment more or less infuscated. Metallic lustre faint. Outer antennal segments (Fig. 231) short and rounded. Pronotum (Fig. 172) narrower and more convex, more constricted at base and usually with sides more rounded, greatest width about middle. Elytra likewise more convex, notably at apex, striae usually finer and with smaller punctures, intervals flatter than in normal *grapii*. Microsculpture entirely lacking on pronotum, on the elytra consisting of isodiametric or quite slightly transverse meshes, in the ♂ visible at extreme apex only. Wings constantly reduced in Fennoscandian specimens, dimorphic in specimens from Siberia and N. America. Penis (Fig. 224) with little developed internal armature.

Distribution. Not in Denmark. — Sweden: recorded only from a restricted area in the extreme northwest (Lule and Torne Lappmark). Very rare. — Norway: very rare: only two isolated records from HO and N. — Finland: only found once in LkW: Pallastunturi. — No records from the adjacent Soviet areas. — Almost circumpolar, in North America only west of the Hudson Bay and in the Rocky Mts.

Biology. Confined to the birch zone and the lower alpine region of the mountains. Habitat almost as in *B. grapii*, but on drier soil, mostly stone-mixed fine sand with sparse vegetation. The beetles occur under stones, among dry grass, etc. Mainly in summer.

154. *Bembidion stephensi* Crotch, 1869
Fig. 225.

Peryphus affinis Stephens, 1835, Ill. Brit. Ent. Mand. 5: 386; *nec* (Say, 1823).
Bembidion stephensi Crotch, 1869, Petites Nouv. Ent. 1: 2.
Bembidium heterocerum Thomson, 1870, Opusc. Ent. 3: 290.

5.2-6.1 mm. Reminding of a large *nitidulum* but paler and with different microsculpture. Upper surface greenish or bluish (elytra sometimes rufinistic). The 3 basal antennal segments, entire palpi, and legs rufo-testaceous. Elytra more oviform, wider in posterior half. Microsculpture transverse without formation of meshes over entire elytra. Penis (Fig. 225) with hypertrophic internal structures.

Distribution. Denmark: scattered in the eastern parts of Jutland (SJ + EJ), F, LFM, SZ, NEZ and B. Also an isolated locality in NWJ: Bulbjerg. Very rarely on inland localities. — Sweden: very rare and with a very limited distribution. Known only from 9 localities in Sk. and one in Vg. (Göteborg). — Norway: very rare, only a few old records from AK (Oslo). — Finland: found once in St: Nakkila and a few times near the south-east border, Ka: Imatra and Sa: Joutseno. — USSR: recorded in Vib: Jääski (= Lesogorskij), several specimens. — Mainly in C. Europe, south to N. Italy.

Biology. On moist clayish or clay-mixed sandy slopes with out-welling water, predominantly on sea slopes, but also along rivers, brooks, etc. It occurs on barren soil or among sparse vegetation of e.g. *Tussilago* and *Equisetum*, usually in somewhat shaded sites, as under bushes. Often associated with *B. nitidulum, genei* and *Stenus fossulatus* Er. During daytime the beetles mostly hide under leaves of *Tussilago*, in cracks, etc. Breeding occurs in spring.

155. *Bembidion lunatum* (Duftschmid, 1812)
Pl. 5: 13.

Elaphrus lunatus Duftschmid, 1812, Fauna Austriae 2: 211.

5.5-6.2 mm. The only species of subgenus *Peryphus* in our fauna with pale spots only in posterior half of elytra. (There are several in Siberia and N. America.) Piceous brown or almost black, upper surface with a bronze hue; elytra with large rufotestaceous semilunar macula near apex (rarely indistinct in specimens with a pale ground-colour). Appendages testaceous or antennae somewhat infuscated apically. Pronotum as in *tetracolum*. 7th elytral stria virtually obsolete. Microsculpture of elytra forming transverse meshes.

Distribution. Denmark: very scattered and rare; SJ (Højer, Ballum sluse), WJ (Fanø, Esbjerg), EJ (Kolding fjord, Vejle fjord, Horsens fjord, Randers fjord), F (Odense, Næsbyhoved). — Sweden: very scattered and rare, found in a few localities (Sk., Bl., Öl., Vg., Vrm., Med. and Jmt.). Most of the specimens have been collected at the river Klarälven. In southern Sweden not established and more or less accidental.

— Norway: scattered in the south-eastern and central districts; also a few local records from N. — Not in East Fennoscandia. — Europe, Siberia, Mongolia.

Biology. On moist clayey or silty ground with rather dense vegetation of grasses, *Juncus, Scirpus*, etc. and often somewhat shaded by scattered bushes. In central and northern Fennoscandia preferably on river banks. In South Scandinavia more common on seashores, e.g. in salt marshes; also in clay pits, etc. Newly emerged beetles usually occur in June. Egg-laying takes place in July-September; larvae hibernate.

156. *Bembidion tetracolum* Say, 1823
Figs 173, 232.

Bembidium tetracolum Say, 1823, Trans. Amer. Philos. Soc. 2: 89.
Bembidion ustulatum auctt.; *nec* (Linnaeus, 1758).
Bembidion litorale auctt.; *nec* (Olivier, 1791).
Bembidium Andreae Thomson, 1859, Skand. Col. 1: 202; *nec* (Fabricius, 1787).

4.9-6.1 mm. This and all following species (except *decorum*) have two pale spots on each elytron, one at base and one near apex; sometimes they are confluent (especially in *obscurellum*). A stout species with a broad pronotum (Fig. 173), and oviform elytra. Pale parts of elytra pronouncedly reddish. Upper surface with faint aeneous lustre. Appendages pale, except that the antennae are infuscated from 3rd to 4th segment. Elytral spots not confluent. Inner elytral striae deep, strongly punctate, 7th stria evident in basal third (as a row of punctures). Pronotum without microsculpture on disc. Microsculpture on the elytra consisting of rather dense, transverse lines, fusing into strongly transverse meshes. Penis, Fig. 232. Wings usually reduced, though with reflexed apex.

Distribution. Denmark: very distributed and common. — Sweden: generally distributed from Sk. to Med. (not G.Sand. and Gstr.) and common south of 60° N. — Norway: mainly confined to coastal areas in the south and south-west. Isolated records also from ST and inner parts of O. — Finland: fairly common in the south, found north to Oa and Kb. — USSR: Vib and southern Kr. — Europe and Siberia; introduced in North America.

Biology. Very eurytopic, occurring on rather moist, usually clayey soil, both in shore habitats, notably along eutrophic waters, and far from shores. Predominantly in open country with scattered vegetation of grasses and weeds, but also in shaded sites: under bushes, in light forest, etc. Often in cultivated fields, where the species may prey upon eggs of insect pests. Breeding occurs in spring.

157. *Bembidion bruxellense* Wesmael, 1835
Figs 233, 238; pl. 5: 11.

Bembidion rupestre auctt.; *nec* (Linnaeus, 1767).

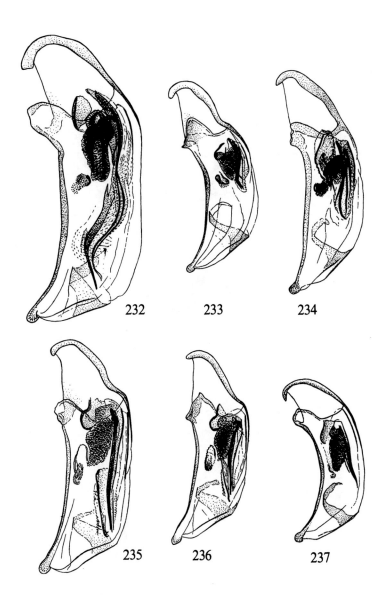

Figs 232-237. Penis of *Bembidion,* subg. *Peryphus.* — 232: *tetracolum* Say; 233: *bruxellense* Wesm.; 234: *petrosum* Gebl.; 235: *andreae* (F.); 236: *femoratum* Sturm; 237: *obscurellum* (Mtsch.).

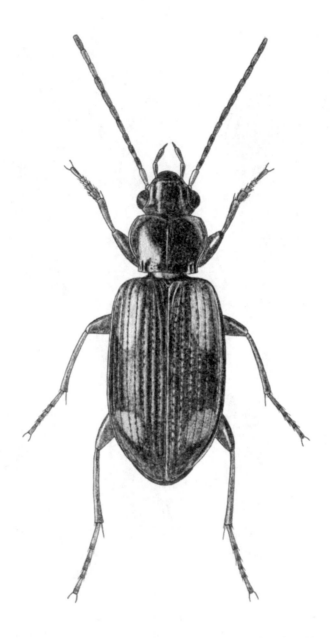

Fig. 238. *Bembidion bruxellense* Wesm., length 4-5.2 mm. (After Victor Hansen).

194

Bembidion bruxellense Wesmael, 1835, Bull. Acad. R. Belg. Cl. Sci., p. 47.

4.0-5.2 mm. Variable in size and extension of the elytral spots, but smaller and darker than *tetracolum*; easily characterized by the dull, transversely microsculptured pronotum (as on the elytra). Forebody more or less aeneous. Second antennal segment (at least apically) and femora infuscated. Elytral pale spots exceptionally so indistinct that they may be overlooked (except in transparent light). Elytra often with a bluish hue, without evident microsculptural meshes. Striae varying and sometimes weaker than in *tetracolum*. Penis, Fig. 233. Wings always full.

Distribution. Denmark: rather distributed and common. — Sweden: generally distributed over the entire country. — Norway: widely distributed and common throughout the country. — Finland: common throughout the country. — USSR: over the entire area covered, north to the Arctic Ocean. — Europe and Siberia; introduced in North America.

Biology. Very eurytopic. On all kinds of moist soil: sand, peat, clay, gravel, etc., with not too dense vegetation. It is most predominant on river banks and along oligo- and dystrophic lakes and ponds, but occurs also far from water, e.g. on waste land, in gravel pits, etc. A spring breeder.

158. *Bembidion petrosum* Gebler, 1833
Figs 175, 234.

Bembidium petrosum Gebler, 1833, Bull. Soc. Nat. Moscou 6: 275.
Bembidion siebkei Sparre-Schneider, 1910, Tromsø Mus. Aarsh. 30: 72.
Bembidion petrosum carlhlindrothi Kangas, 1980, Ent. Gener. 6: 364.

5.0-5.7 mm. Superficially imitating *bruxellense* but differing in the following points: slenderer with longer, more parallel-sided elytra. Forebody with a rather pronounced aeneous tinge, elytra often faintly bronzed, their ground-colour often brownish and the anterior pale spot therefore sometimes not clearly delimited, the posterior reaching apex. Three basal segments of antennae and legs entirely pale rufous, or rarely tip of 3rd segment and femora very slightly darkened (less than in *bruxellense*). Antennae slenderer, notably outer segments. Pronotum narrower, base longitudinally rugulose. Elytral striae somewhat varying but usually shallower and with finer punctures than in the two preceding species. Outer striae usually less obsolete at apex, 7th always with evident punctures behind shoulder. Microsculpture lacking on centre of pronotum on the elytra as in *tetracolum* but meshes usually shorter (notably in the ♀), about twice as wide as long. Penis, Fig. 234. Wings always full.

The Fennoscandian form has been regarded as a ssp. (*"siebkei"* (Sparre-Schneider) J. Müller), but the small differences from the Siberian *petrosum* s.str. are unimportant and overbridged by specimens from other areas, e.g. in N. America (Lindroth, 1963: 334) and not necessary to maintain.

Distribution. Not in Denmark. — Sweden: very rare. Only known from two localities (Vrm., Fastnäs and (T.Lpm., Abisko). — Norway: rather common in N, TR and F. A few scattered records from NT, ST, MR, O and southern HE; can locally be rather common. — Finland: many records from Li: Utsjoki, also found in Ks: Kuusamo. — USSR: recorded from southern Kr ("ssp. *carlhlindrothi*") and from northernmost Kr. — Circumpolar.

Biology. Almost confined to river banks, rarely on lake shores, inhabiting moist, exposed sites without or with sparse vegetation. It has a clear preference for silty ground covered by gravel or boulders (Andersen 1970, 1978); less often occurring on fine sand. Oviposition takes mainly place in June-July; imagines, but now and then also larvae or pupae, hibernate.

159. *Bembidion andreae* (Fabricius, 1787)
Fig. 235.

Carabus Andreae Fabricius, 1787, Mant. Ins. 1: 204.
Bembidium cruciatum Schiødte, 1841, Gen. Spec. Danm. Eleuth. 1: 337; *nec* Dejean, 1831.
Bembidium concinnum Thomson, 1871, Opusc. Ent. 4: 361; *nec* Stephens, 1828.
Bembidion andreae polonicum J. Müller, 1930, Col. Centralbl. 5: 1.
Bembidion femoratum var *dissolutum* Hellén, 1934, Notul. Ent. 14: 55.

4.5-5.5 mm. Flatter than *tetracolum*. Pale parts more yellow. Elytral striae finer, 7th barely suggested. Black with a faint aeneous hue, notably on forebody. Antennae with 3 basal segments, maxillary palpi (except penultimate segment) and entire legs pale (sometimes femora very slightly infuscated). The pale elytral spots sharply delimited, the anterior usually both inwards and posteriorly more extended than in the three preceding species and may be confluent with the posterior spot. Microsculpture of elytra stronger than in *bruxellense* but without formation of meshes (as in *petrosum*). Penis, Fig. 235. Wings full.

B. andreae is a complex species and our form has been referred to ssp. *polonicum* J. Müller (*dissolutum* Hellén). The nominate ssp. belongs to the W. Mediterranean.

Distribution. Denmark: rare and very scattered, on coastal localities in all districts except WJ, NWJ and NWZ. — Sweden: a few localities on the west coast of Skåne and Halland; also Gtl., Fårön, 2 specimens 1901. Rather common on the island Ven in Öresund and at Ålabodarna in western Skåne; elsewhere extremely rare. — Not in Norway. — Finland: a few localities along the southern coast. — USSR: common in Vib, mainly at the coast; in Kr found north to Karhumäki. — Europe, western Siberia and the Caucasus.

Biology. On moist, barren or sparsely vegetated clay-mixed sand near water. In Sweden and Denmark confined to seashores, often occurring on clayey sea slopes with trickling water; in Finland exclusively on sandy river banks. Breeding occurs in spring.

160. *Bembidion femoratum* Sturm, 1825
Fig. 236; pl. 5: 12.

Bembidium femoratum Sturm, 1825, Deutschl. Fauna Ins. 5 (6): 117.

4.2-5.2 mm. Closely related to *andreae* and has often been regarded as a subspecies of this. Darker and usually smaller. Only 1 or 2 basal segments of antennae entirely pale, also ante-penultimate segment of maxillary palpi infuscated, femora almost black. Forebody without or with a faint metallic hue. Pronotum with less rounded sides, the basal punctuation often confluent into longitudinal wrinkles. Legs shorter. Microsculpture as in *andreae*. Penis (Fig. 236) shorter, especially the sclerites of the inner armature. Wings full.

Distribution. Denmark: distributed and rather common in SJ, EJ and all the eastern provinces (F - B). Also a few localities in NWJ at the Limfjord and in NEJ near the Skaw. — Sweden: generally distributed in the south up to Med. and Jmt. and rather common. Scattered distributed and rare in Ång., Vb., Nb., Ly.Lpm. and P.Lpm. — Norway: scattered, but widely distributed and locally common in most districts. — Finland: scattered but widely distributed, in the north to Li: Utsjoki. — USSR: scattered records from Vib and southern Kr. — Europe, Siberia and Asia Minor.

Biology. Moderately hygrophilous, occurring mainly on clay or clay-mixed sand, also on silt or gravel, usually among sparse vegetation. Predominantly near water, notably on river banks, less often on lake- and seashores. Frequently also far from water, e.g. in gravel and clay pits, at road sides and in cultivated fields. Reproduction takes place in spring.

161. *Bembidion maritimum* (Stephens, 1835)

Peryphus maritimus Stephens, 1835, Ill. Brit. Ent. Mand. 5: 385.

5.0-5.5 mm. Easily recognized on the elytral pattern and the microsculpture. Forebody black with a greenish lustre. The pale elytral spots are confluent along the side-margin. Appendages entirely testaceous or antennae slightly infuscated apically. Elytra stretched, parallel-sided, with strongly punctate striae. Their microsculpture is strong, more or less isodiametric in the ♀, in the ♂ consisting of meshes about twice as broad as long.

Distribution. In our area only in SW. Denmark: several localities in SJ and WJ from the German border north to Skallingen. — W. Europe.

Biology. Very stenotopic, confined to salt marshes on the seashore and to river estuaries. On the Danish North Sea coast it occurs in the tidal zone, on heavy marine clay with more or less developed vegetation, together with, among others, *Pogonus chalceus* and *luridipennis*. Under stones and seaweed, in clay cracks, etc. Predominantly in spring.

162. *Bembidion obscurellum* (Motschulsky, 1845)
Figs 174, 237; pl. 5: 16.

Peryphus obscurellus Motschulsky, 1845, Bull. Soc. Nat. Moscou 18 (1): 27.
Bembidium repandum J. Sahlberg, 1875, Notis. Sällsk. Fauna Flora Fenn. Förh. 14: 78.

3.9-5.1 mm. Small and short. Best recognized on the microsculpture. Black, forebody aeneous or greenish, pale elytral maculae so expanded that normally only 1st interval, extreme side-margin, and a triangular spot across the suture behind middle remain dark (brown or almost black); in the darkest specimens the central spot reaches side-margin, in the palest it is only suggested. Appendages normally entirely rufo-testaceous but 2nd segment of maxillary palpi, antennae from tip of 4th segment and the femora may be slightly infuscated. Pronotum (Fig. 174) with constricted basal portion very short. Elytra with striae finely punctate, 7th stria irregular or virtually absent. Microsculpture wanting on centre of pronotum, strong on the elytra (notably in the ♀), consisting of isodiametric or very slightly transverse meshes without transverse arrangement. ♂ with armature of internal sac shorter than in all related species (Fig. 237). Wings full.

Distribution. Denmark: accidental finds in WJ (Blåvand in 1971, Vejrs in 1931) and NEJ (Skagen in 1915). — Sweden: only occasionally found in Sweden, the finds undoubtedly representing drifted specimens. Collected exclusively on seashores. The first specimen was found 1972 (Sk., Sandhammaren). Other finds are Sk., Sandhammaren (1975); Sk., Vitemölla (1975); Gotska Sandön (1975); Gtl., Fårön, Sudersand (1981); Gtl., Sandviken S Herrvik (1981). — Not in Norway or Finland. — In the USSR several records from the south coast of the Kola peninsula. — Circumpolar.

Biology. On dry sandy soil, not dependent of water, though in our area restricted to sandy seashores. The Scandinavian specimens are undoubtedly accidental stragglers, washed ashore. Baranowski (1976) assumes that a small but vigorous population exists somewhere at the Baltic Sea.

163. *Bembidion saxatile* Gyllenhal, 1827

Bembidion saxatile Gyllenhal, 1827, Ins. Suec. 4: 406.

4.2-5.1 mm. Very flat, with long parallel-sided elytra. Best recognized on a group of small punctures (as in *decorum*) on frons inside posterior part of eye. Forebody green, elytra with a bluish hue, the pale spots clear reddish. Legs entirely pale rufous or with faintly infuscated femora; as a rule only 1st antennal segment entirely rufous. Pronotum strongly dilated anteriorly. Elytral stria evident to apex. Microsculpture as in *femoratum*.

Distribution. Denmark: in Jutland scattered distributed along the eastern coasts of SJ and EJ north to Mols; further around central parts of the Limfjord (NWJ and NEJ) and formerly also at the North Sea coast of NEJ. Scattered distributed in all the eastern districts (F - B). — Sweden: recorded from all districts except P.Lpm. and T.Lpm. Rather common and distributed in the south to 63° N, north of 63° N scattered distributed, apparently lacking over large areas. — Norway: scattered throughout the country, but common only in the south. Rare in F. — Finland: widely distributed and locally common, recorded north to Li: Utsjoki. — USSR: scattered over Vib, Kr and Lr. — Europe, Siberia and the Caucasus.

Biology. Predominantly on barren, gravelly and stony banks and shores of rivers and lakes; also on seashores, and always near the water edge. In the Fennoscandian mountains rarely above the coniferous zone. In Denmark the species in confined to clayey sea slopes, often with trickling water, usually occurring among gravel and stones at the bottom of the slopes. During daytime the beetles are mostly found in concealment under stones, etc. Breeding occurs in spring.

164. *Bembidion decorum* (Zenker, 1801)

Carabus decorus Zenker, 1801, *in* Panzer: Faun. Ins. Germ. 73: 4.

5.2-6.0 mm. Piceous black, with an aeneous tinge, elytra often rufinistic. 1st antennal segment and legs rufo-testaceous. Lateral frontal punctures inside eyes rather strong. Pronotum without latero-basal carina. Elytral striae strongly punctate anteriorly, the inner deepened but obliterating before apex. Microsculpture consisting of dense, confluent transverse lines.

Distribution. The only record from our area is Denmark: SJ, Haderslev, in great numbers in May 1881. — C. and S. Europe, the Mediterranean area.

Biology. On gravelly banks of rivers.

Genus *Tachys* Dejean, 1821

Tachys Dejean, 1821, Cat. Coll. Col. B. Dejean: 16.
  Type-species: *Tachys scutellaris* Stephens, 1828.
Elaphropus Motschulsky, 1839, Bull. Soc. Nat. Moscou 12: 73.
  Type-species: *Elaphropus caraboides* Motschulsky, 1862.
Tachyura Motschulsky, 1862, Etud. Ent. 11: 27.
  Type-species: *Elaphrus quadrisignatus* Duftschmid, 1812.
Porotachys Netolitzky, 1914, Ent. Blätt. 10: 174.
  Type-species: *Trechus bisulcatus* Nicolai, 1822.
Paratachys Casey, 1918, Mem. Coleopt. 8: 174.
  Type-species: *Paratachys austinicus* Casey, 1918.
Eotachys Jeannel, 1941, Faune de France 39: 426.
  Type-species: *Elaphrus bistriatus* Duftschmid, 1812.

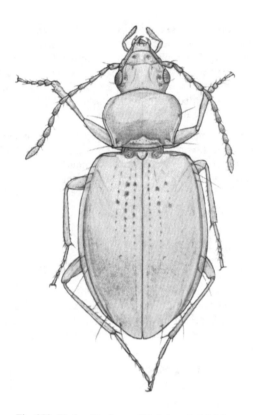

Fig. 239. *Tachys bisulcatus* (Nic.), length 2.8-3.2 mm.

The genus has recently been divided into a number of genera. They are very abundant in warmer regions, substituting *Bembidion*. In these divisions the North European species go into different groups. As divided by Erwin (1974) *T. bistriatus* is included in the genus *Paratachys*, of which *Eotachys* is a junior synonym, *T. bisulcatus* is included in the genus *Porotachys*, and *T. parvulus* is included in *Tachyura*, which is considered a subgenus of *Elaphropus*. Often *Paratachys* is considered a subgenus of the genus *Tachys* in its more restricted sense.

Very small species, less than 3.2 mm long, agreeing with *Bembidion* in the reduction of the last segment of the maxillary palpi. Easily separated from that genus (except from subgenus *Ocys*) on the recurrent sutural stria of elytra (Figs 243, 244), as in *Trechus*, and the obliquely truncate tip of the pro-tibia. Legs and at least base of antenna pale. Second antennal segment slender, about as long as third. Elytra without abbreviated scutellar stria, the outer striae obliterated; 2 dorsal punctures, situated at

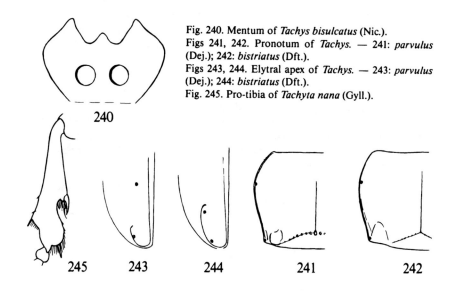

Fig. 240. Mentum of *Tachys bisulcatus* (Nic.).
Figs 241, 242. Pronotum of *Tachys.* — 241: *parvulus* (Dej.); 242: *bistriatus* (Dft.).
Figs 243, 244. Elytral apex of *Tachys.* — 243: *parvulus* (Dej.); 244: *bistriatus* (Dft.).
Fig. 245. Pro-tibia of *Tachyta nana* (Gyll.).

240

245   243   244   241   242

third stria or on fourth interval, the posterior often enclosed within recurrent stria; behind these a subapical puncture (Figs 243, 244). The wings are full in our species. Male pro-tarsi with 2 faintly dilated basal segments (in our species).

Key to species of *Tachys*

1　Posterior dorsal puncture of elytra situated well in front of recurrent stria (Fig. 243). Convex species with shiny upper surface, microsculpture absent or very fine 2
-　Posterior dorsal puncture enclosed within the hook of the recurrent stria (Fig. 244). Flat species with upper surface dull from dense transverse microsculpture, causing iridenscence . . . . 165. *bistriatus* (Duftschmid)

2(1)　Unicolorous rufo-testaceous. Elytra strongly dilated at middle, inner striae with coarse punctures (Fig. 239) . . . . . 166. *bisulcatus* (Nicolai)
-　Darker. Elytra almost parallel-sided; their striae almost impunctate . . . . 167. *parvulus* (Dejean)

165. *Tachys bistriatus* (Duftschmid, 1812)
　　Figs 242, 244.

Elaphrus bistriatus Duftschmid, 1812, Fauna Austriae 2: 205.

1.8-2.3 mm. A flat weakly sclerotized species. Upper surface dull and more or less iridescent, with dense transverse microsculpture. Somewhat reminiscent of *Trechus quadristriatus*, but much smaller. Piceous to brown, head darkest, antennae with testaceous base. Pronotum (Fig. 242) with sides sinuate before hind-angles which are obtuse; transverse basal impression faint and impunctate. Elytral striae faint, impunctate. Male with 2 dilated pro-tarsal segments.

Distribution. Very rare in Denmark: EJ, Samsø, 1 specimen July 1981; F, Ærø, in number 1932; Ristinge (Langeland), 1942 and since; Helnæs, in number 1962; B, Hasle, 1 specimen. — Sweden: a single specimen found in Ög., Vist 6. iv. 1967; certainly accidental and probably not established. — Not in Norway. — Finland: only known from Kb, Kitee (several specimens). More common in Kr around Lake Ladoga in the USSR. — Europe, North Africa, the Canaries, the Caucasus.

Biology. On damp sand and clay at the border of standing or running freshwater, both inland and at the coast. In Denmark it is confined to clayey sea-slopes, occurring under sea drift and in clay cracks. In Karelia it has been encountered numerously in wood waste deposits on a river bank. Mainly in spring.

166. *Tachys bisulcatus* (Nicolai, 1822)
Figs 239, 240; pl. 5: 21.

Trechus bisulcatus Nicolai, 1822, Diss. Col. Agr., p. 26.
Tachys frontalis Hayward, 1900, Trans. Am. Ent. Soc. 26: 212.

2.8-3.2 mm. Rufo-testaceous with dark eyes. Mentum with two deep, round foveae. Pronotum with a small punctiform fovea in front of the rectangular hind-angle, but without latero-basal carina; transverse basal impression punctate. Elytra very broad, oviform, convex, with 8th stria obsolete at middle. Pronotum and elytra with extremely dense and fine transverse microsculptue and therefore faintly iridescent. Pro-tarsi of male with 2 sligthly dilated basal segments.

Distribution. Denmark: very rare, recorded for the first time in 1969 and since apparently spreading. SJ (Stensbæk), WJ (Billund), EJ (Ejstrupholm, Samsø), F (Tranekær), LFM (Bøtø), NEZ (St. Hareskov, Stampen). — Sweden: rare and local but found more frequently in the last years. Reported for the first time in 1928, now known from many districts from Sk. to T.Lp.. Common in Skåne in 1971. — Norway: only a few records: AK, HEs, NTi. — Finland: widespread but rare, northernmost find in Li: Inari 1972. — USSR: Vib and southern Kr. — A species in rapid expansion in Europe, found also in North Africa and introduced in North America.

Biology. Originally associated with coniferous forest, now usually found at sawmills in heaps of damp, fermenting spruce bark, where the development is undertaken. Also at garbage deposits. The species is often observed swarming at night or at sunset. It is most numerous in spring, when propagation takes place. Young adults emerge in autumn.

167. *Tachys parvulus* (Dejean, 1831)
  Figs 241, 243.

Bembidium parvulum Dejean, 1831, Spec. Gén. Col. 5: 57.

1.8-2.2 mm. Upper surface devoid of microsculpture, very shiny. Piceous brown to almost black, elytra often a little paler. Basal impression of pronotum deep with strong punctures; latero-basal fovea distinct. 8th elytral stria entire. Male with 2 slightly dilated pro-tarsal segments.

Distribution. Only Sweden: Hall., Träslöv, 1 specimen 12.x.1962; it is unknown whether a permanent population exists here. — C. Europe, S. England; introduced in North America.

Biology. It occurs on open ground, often near the sea.

Genus *Tachyta* Kirby, 1837

Tachyta Kirby, 1837, Fauna Bor. - Amer. 4: 56.
  Type-species: *Tachyta picipes* Kirby (= *nana* Gyllenhal).

Earlier regarded as a subgenus of *Tachys*, but better treated as a distinct genus. It is particularly characterized by the recurrent elytral stria running close to and almost parallel with the side-margin. 8th stria is complete as in *Tachys parvulus*. The anterior dorsal puncture is situated closer to base than to suture. The antennae are very short. The body is very flat and carries unusually long setae, an adaptation to a life under bark. Upper surface dull from very coarse, irregularly isodiametric microsculpture. A single species in Europe.

168. *Tachyta nana* (Gyllenhal, 1810)
  Fig. 245; pl. 5: 22.

Bembidium nanum Gyllenhal, 1810, Ins. Suec. 2: 30.

2.6-3.2 mm. Black to piceous, elytra sometimes slightly paler, notably along sides and suture, as are all appendages (especially base of antennae). Frontal furrows well delimited. Pronotum with a weak latero-basal carina. Striation of elytra highly varying, from only 3 to 6 clearly visible. Male with 2 dilated pro-tarsal segments. Internal sac of penis with complex of sclerotized structures.

Distribution. Not in Denmark. — Sweden: rather common in northern and central Sweden (Sdm. - Lu.Lpm., except Hrj.). Very rare in Sk., Sm. and Ög. and seemingly decreasing in number in this area. — Norway: Ø, AK, HE, Bø, TE, TRi, Fø. — Finland: found in most of the country (except Al) but not common. — A circumpolar species, mainly in the north.

Biology. Under loose bark of tree stumps and dead trunks, in sun-exposed situations. It prefers coniferous trees, especially pine, but is also regularly found associated with birch (e.g. in Norway), rarely with other deciduous trees. Both larva and imago attack species of bark beetles and their progeny. Reproduction takes place in spring; young beetles occur in autumn.

Tribe Pogonini

A small group of species reminding of a large *Bembidion* in general habitus, but with last palpal segment fully developed.

Confined to the sea-coast or inland saline localities. Only one genus in our fauna.

Genus *Pogonus* Dejean, 1821

Pogonus Dejean, 1821, Cat. Coll. Col. B. Dejean: 9.
Type-species: *Carabus littoralis* Duftschmid, 1812.

Medium-sized metallic species. Raised basal margin of elytra complete; 3 dorsal punctures. Frontal furrows deep and straight. Base of pronotum punctate. Tarsi furrowed on dorsum. Wings full. Male with 2 dilated pro-tarsal segments.

Key to species of *Pogonus*

1　Elytra yellowish brown. Appendages rufo-testaceous　169. *luridipennis* (Germar)
-　Elytra piceous to black. Appendages darkened . . . . 170. *chalceus* (Marsham)

169. *Pogonus luridipennis* (Germar, 1822)
　　　Fig. 246.

Harpalus luridipennis Germar, 1822, Fauna Ins. Eur. 7: 3.

6-8.5 mm. At once recognized by the pale testaceous elytra, but sometimes clouded on the disc and/or with a faint metallic hue. Forebody green. All appendages rufo-testaceous. Pronotum flatter than in *chalceus*, and elytral striae stronger apically.

Distribution. Denmark: very rare, only in the extreme south-western part of Jutland, SJ (Ballum sluse) and WJ (Esbjerg, Ho bugt). — Sweden: the only record is from Boh., Solberga, where it was discovered 1943. A permanent and very isolated population still exists here. — Norway: very rare, only Ø. — Not in East Fennoscandia. — Along seashores of W. Europe and Asia, in C. Europe also in saline places.

Biology. A halobiontic beetle, living in salt marshes on moist clay soil with scattered vegetation of e.g. *Salicornia*, usually in habitats that are regularly flooded by the tide. In C. Europe also found on inland saline localities. It occurs among plant roots, under

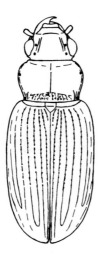

Fig. 246. *Pogonus luridipennis* (Germ.).

seaweed and in clay cracks, often together with *P. chalceus, Bembidion minimum* and *Dicheirotrichus gustavi*. The species is most numerous in May-June, which is the breeding period, and in the autumn when the young adults emerge.

170. *Pogonus chalceus* (Marsham, 1802)

Carabus chalceus Marsham, 1802, Ent. Brit., p. 460.

5.5-6.5 mm. Entire upper surface metallic: bronze, brassy, or greenish; elytra concolorous. At least 1st antennal segment black or piceous; also palpi and legs more or less infuscated. Pronotum more convex and with more rounded sides. Elytra shorter, more oviform. Striae finer, obsolete both laterally and apically.

Distribution. Denmark: distributed along the SW. coast of Jutland (SJ and WJ), north to Skallingen. Also in EJ: Stavns fjord on Samsø, 1 specimen 1977; F: Ristinge on Langeland, in number before 1900. — Not in Sweden, Norway or Finland.

Biology. A halobiontic species, living in the same habitats as *P. luridipennis*; the two species are often found together along the North Sea coast of S. Jutland. It is especially numerous in newly diked areas (Heydemann 1962); in C. Europe also occurring in saline inland localities. Mainly met with in spring and summer, when the eggs are laid, and again in the autumn, when the newly emerged beetles occur.

		Germany	G. Britain	SJ	EJ	WJ	NWJ	NEJ	F	LFM	SZ	NWZ	NEZ	B	Sk.	Bl.
Cicindela sylvatica L.	1	●	●	●	●	●	●	●	●	●		●	●	●	●	●
C. hybrida L.	2	●	●	●	●	●	●	●	●	●	●	●	●	●	●	●
C. maritima Latr. & Dej.	3	●	●	●	●	●		●		●			●	●		●
C. campestris L.	4	●	●	●	●	●	●	●	●	●	●	●	●	●	●	●
Trachypachus zetterstedti (Gyll.)	5															
Omophron limbatum (F.)	6	●	●	●	●				●	●	●	●	●	●		
Calosoma sycophanta (L.)	7	●	●	●	●				●	●	●	●	●			
C. inquisitor (L.)	8	●	●	●	●		●		●	●	●	●			●	●
C. m. auropunctatum (Hbst.)	9	●		●	●			●		●			●		●	
C. denticolle Gebl.	10															
C. investigator (Ill.)	11															
C. reticulatum (F.)	12	●		●	●										●	
Carabus monilis F.	13	●	●													
C. arvensis Hbst.	14	●	●	●	●	●	●	●			●	●	●			
C. granulatus L.	15	●	●	●	●	●	●	●	●	●	●	●	●	●	●	●
C. menetriesi Hummel	16	●														
C. clathratus L.	17	●	●	●	●	●	●	●	●	●	●		●		●	●
C. cancellatus Ill.	18	●	●	●	●	●		●	●	●	●	●			●	●
C. auratus L.	19	●	●						●				●		●	
C. nemoralis Müller	20	●	●	●	●	●	●	●	●	●	●	●	●	●	●	●
C. hortensis L.	21	●	●	●	●	●	●	●	●	●	●	●	●	●	●	●
C. glabratus Payk.	22	●	●	●	●			●					●		●	●
C. problematicus gallicus Géhin	23a	●	●	●	●		●	●							●	●
C. p. wockei Born	23b															
C. p. strandi Born	23c															
C. nitens F.	24	●	●	●	●	●	●	●			●	●	●			●
C. convexus F.	25	●		●	●				●	●	●	●	●			
C. intricatus L.	26	●	●		●								●	●		
C. violaceus L.	27	●	●	●	●	●	●	●	●	●	●	●	●	●		●
C. coriaceus L.	28	●	●	●	●	●	●	●	●	●	●	●	●	●		●
Cychrus caraboides (L.)	29	●	●	●	●	●	●	●	●	●	●	●	●	●		●
Leistus rufomarginatus (Dft.)	30	●	●	●	●	●	●	●	●	●	●	●	●	●		●
L. terminatus (Hellw. in Pz.)	31	●	●	●	●	●	●	●	●	●	●	●	●	●		●
L. ferrugineus (L.)	32	●	●	●	●	●	●	●	●	●	●	●	●	●	●	●
Nebria livida (L.)	33	●	●	●	●	●	●	●	●		●	●	●	●		●
N. rufescens (Ström)	34	●	●													
N. nivalis (Payk.)	35	●														
N. brevicollis (F.)	36	●	●	●	●	●	●	●	●	●	●	●	●	●	●	●

	Hall.	Sm.	Öl.	Gtl.	G. Sand.	Ög.	Vg.	Boh.	Dlsl.	Nrk.	Sdm.	Upl.	Vstm.	Vrm.	Dlr.	Gstr.	Hls.	Med.	Hrj.	Jmt.	Ång.	Vb.	Nb.	Ås. Lpm.	Ly. Lpm.	P. Lpm.	Lu. Lpm.	T. Lpm.
1	●	●	●	●	●	●	●	●	●	●	●	●	●	●	●	●	●	●	●	●	●	●	●	●	●	●	●	●
2	●	●				●	●	●	●					●														
3	●														●	●		●		●	●	●	●					
4	●	●	●	●		●	●	●	●	●	●	●	●	●	●	●	●	●	●	●	●	●			●			
5																	●					●					●	●
6	●	●				●	●																					
7		●	●				●	●						●														
8	●	●	●			●	●	●		●	●	●	●															
9	●		●																									
10																												
11			●																									
12			●						●																			
13																												
14	●	●				●	●	●	●	●	●	●	●	●									●					
15	●	●	●	●		●	●	●	●	●	●	●	●	●						●	●							
16																												
17	●	●	●	●		●	●	●	●	●	●	●	●			●					●	●	●	●	●	●	●	●
18	●	●	●			●	●	●	●	●	●	●	●															
19											●																	
20	●	●	●			●	●	●	●	●	●	●	●	●		●					●	●						
21	●	●	●			●	●	●	●	●	●	●	●	●						●	●	●	●					
22	●	●	●			●	●	●	●	●	●	●	●	●				●	●	●	●	●	●	●	●			
23a	●	●				●	●	●	●	●	●	●	●															
23b														●					●	●								
23c																												●
24	●	●	●	●		●	●	●	●	●	●	●	●	●	●	●	●											
25	●					●	●	●	●																			
26																												
27	●	●	●	●		●	●	●	●	●	●	●	●	●				●	●	●	●	●	●	●	●	●	●	●
28	●	●	●			●	●	●		●	●	●	●	●	●													
29	●	●	●	●	●	●	●	●	●	●	●	●	●	●	●	●	●	●	●	●	●	●	●	●	●	●	●	●
30	●						●																					
31	●	●	●			●	●	●	●	●	●	●	●	●	●	●	●	●		●				●		●	●	
32	●	●	●	●	●	●	●	●	●	●	●	●	●	●	●	●	●	●		●				●			●	
33	●	●	●			●	●		●	●				●														
34		●	●			●	●		●	●		●	●				●	●	●	●	●	●	●	●	●	●	●	●
35																	●								●	●	●	●
36	●	●	●	●	●	●	●	●	●		●			●														

		Ø+AK	HE (s+n)	O (s+n)	B (ø+v)	VE	TE (y+i)	AA (y+i)	VA (y+i)	R (y+i)	HO (y+i)	SF (y+i)	MR (y+i)	ST (y+i)	NT (y+i)	Ns (y+i)
Cicindela sylvatica L.	1	●	●	●	●		●	●	●	●	●		●			
C. hybrida L.	2			●		●										
C. maritima Latr. & Dej.	3		●	●	●								●	●	●	
C. campestris L.	4	●	●	●	●	●	●	●	●	●	●	●	●	●	●	●
Trachypachus zetterstedti (Gyll.)	5															●
Omophron limbatum (F.)	6															
Calosoma sycophanta (L.)	7															
C. inquisitor (L.)	8	●						●	●							
C. m. auropunctatum (Hbst.)	9	●														
C. denticolle Gebl.	10															
C. investigator (Ill.)	11															
C. reticulatum (F.)	12															
Carabus monilis F.	13	●														
C. arvensis Hbst.	14	●	●		●	●	●	●								
C. granulatus L.	15	●	●	●	●	●	●	●	●	●	●	●				
C. menetriesi Hummel	16															
C. clathratus L.	17	●				●					●					
C. cancellatus Ill.	18	●	●		●	●	●	●			●					
C. auratus L.	19	●														
C. nemoralis Müller	20	●	●	●	●	●	●	●	●	●	●	●	●	●		
C. hortensis L.	21	●	●	●	●	●	●	●	●	●	●	●	●	●	●	●
C. glabratus Payk.	22	●	●	●	●	●	●	●	●	●	●	●	●	●	●	●
C. problematicus gallicus Géhin	23a	●									●	●	●	●		
C. p. wockei Born	23b		●	●	●											
C. p. strandi Born	23c															
C. nitens F.	24	●	●	●	●	●	●	●	●	●	●	●		●	●	●
C. convexus F.	25	●	●		●	●		●		●						
C. intricatus L.	26															
C. violaceus L.	27	●	●	●	●	●	●	●	●	●	●	●	●	●	●	●
C. coriaceus L.	28	●	●		●		●	●	●	●	●	●	●	●		
Cychrus caraboides (L.)	29	●	●	●	●	●	●	●	●	●	●	●	●	●	●	●
Leistus rufomarginatus (Dft.)	30															
L. terminatus (Hellw. in Pz.)	31	●	●	●				●	●	●	●	●	●	●	●	
L. ferrugineus (L.)	32	●		●	●	●	●	●	●	●	●	●	●	●	●	
Nebria livida (L.)	33	●			●		●			●						
N. rufescens (Ström)	34	●	●	●	●	●	●	●	●	●	●	●	●	●	●	●
N. nivalis (Payk.)	35		●	●	●						●		●		●	●
N. brevicollis (F.)	36	●				●	●	●	●	●	●	●	●			

	Nn (ø+v)	TR (y+i)	F (v+i)	F (n+ø)	Al	Ab	N	Ka	St	Ta	Sa	Öa	Tb	Sb	Kb	Om	Ok	Ob S	Ob N	Ks	LkW	LkE	Le	Li	Vib	Kr	Lr
1					●	●	●	●	●	●	●	●	●	●	●	●	●	●	●	●	●	●	●	●	●	●	●
2						●	●	●	●	●				●	●	●									●	●	
3			●			●	●	●	●	●						●			●	●	●	●			●	●	●
4		●			●	●	●	●		●	●	●	●	●	●	●			●	●					●		
5		●		●			●	●								●				●	●			●	●	●	●
6																											
7																											
8						●	●																		●		
9																											
10						●																					
11																											
12																											
13																											
14						●	●	●	●	●	●		●												●	●	
15					●	●	●	●	●	●	●	●	●	●	●		●		●						●	●	
16																									●	●	
17						●	●	●	●	●	●	●	●		●	●	●	●	●	●					●	●	
18						●	●	●	●	●	●		●	●		●									●	●	
19																											
20						●	●	●	●	●	●	●	●	●		●									●	●	
21						●	●	●	●	●	●	●	●	●	●										●	●	
22	●	●	●	●	●	●	●	●	●	●	●	●	●	●	●	●	●	●	●	●	●	●	●	●	●	●	●
23a																											
23b																									●		
23c	●	●	●	●																		●	●				●
24	●				●	●	●	●	●	●	●	●	●	●	●	●	●	●	●	●	●	●	●	●	●	●	●
25					●	●	●																		●		
26																											
27	●	●	●		●	●	●		●	●	●		●			●	●	●	●	●	●	●			●	●	●
28																											
29	●	●	●	●	●	●	●	●	●	●	●		●	●	●	●		●	●	●	●				●	●	●
30																											
31		●	●		●	●	●	●	●	●	●	●	●	●	●	●	●	●		●					●	●	
32	●	●	●		●	●	●	●	●			●		●											●	●	
33							●			●	●	●													●	●	
34	●	●	●	●	●	●	●			●	●		●	●		●	●	●	●	●	●	●	●		●	●	●
35	●	●	●	●																	●		●	●			●
36				●																							

209

		Germany	G. Britain	SJ	EJ	WJ	NWJ	NEJ	F	LFM	SZ	NWZ	NEZ	B	Sk.	Bl.
Nebria salina (Fairm. & Lab.)	37	●	●	●	●	●	●	●	●	●	●	●	●	●		●
Pelophila borealis (Payk.)	38	●														
Notiophilus aestuans Mtsch.	39	●	●	●	●	●	●	●	●	●	●	●	●	●		●
N. aquaticus (L.)	40	●	●	●	●	●	●	●	●	●	●	●	●	●	●	●
N. palustris (Dft.)	41	●	●	●	●	●	●	●	●	●	●	●	●	●	●	●
N. germinyi Fauv.	42	●	●	●	●	●	●	●	●	●	●	●	●	●	●	●
N. rufipes Curt.	43	●	●	●	●	●			●	●	●	●		●		
N. reitteri Spaeth	44															
N. biguttatus (F.)	45	●	●	●	●	●	●	●	●	●	●	●	●	●	●	●
Blethisa multipunctata (L.)	46	●	●	●	●	●	●	●	●	●	●	●	●	●	●	●
Diacheila arctica (Gyll.)	47															
D. polita (Fldm.)	48															
Elaphrus lapponicus Gyll.	49															
E. uliginosus F.	50	●	●	●				●	●	●	●	●	●	●	●	●
E. cupreus Dft.	51	●	●	●	●	●	●	●	●	●	●	●	●	●	●	●
E. riparius (L.)	52	●	●	●	●	●	●	●	●	●	●	●	●	●	●	●
E. angusticollis F. Sahlbg.	53															
Loricera pilicornis (F.)	54	●	●	●	●	●	●	●	●	●	●	●	●	●	●	●
Clivina fossor (L.)	55	●	●	●	●	●	●	●	●	●	●	●	●	●	●	●
C. collaris (Hbst.)	56	●	●	●	●								●		●	●
Dyschirius thoracicus (Rossi)	57	●	●	●	●	●	●	●	●	●	●	●	●	●	●	●
D. obscurus (Gyll.)	58	●	●	●	●	●	●	●	●	●	●	●	●	●		
D. angustatus (Ahr.)	59	●	●	●	●	●		●	●				●		●	
D. nitidus (Dej.)	60	●	●													
D. neresheimeri Wagn.	61	●														
D. politus (Dej.)	62	●	●	●	●	●	●	●	●	●	●	●	●	●	●	●
D. impunctipennis Daws.	63	●	●	●	●	●		●		●		●		●	●	
D. chalceus Er.	64	●			●		●		●			●		●		●
D. salinus Schaum	65	●	●	●	●	●	●	●	●	●	●	●	●	●		
D. aeneus (Dej.)	66	●	●	●				●		●	●	●	●	●		
D. luedersi Wagn.	67	●	●	●	●	●	●	●	●	●	●	●	●	●		●
D. septentrionum Munst.	68															
D. nigricornis Mtsch.	69															
D. intermedius Putz.	70	●		●	●	●	●			●		●	●		●	●
D. laeviusculus Putz.	71														●	
D. globosus (Hbst.)	72	●	●	●	●	●	●	●	●	●	●	●	●	●	●	●
Broscus cephalotes (L.)	73	●	●	●	●	●	●	●	●	●	●	●	●	●	●	●
Miscodera arctica (Payk.)	74	●	●		●			●					●	●	●	●

	Hall.	Sm.	Öl.	Gtl.	G. Sand.	Ög.	Vg.	Boh.	Dlsl.	Nrk.	Sdm.	Upl.	Vstm.	Vrm.	Dlr.	Gstr.	Hls.	Med.	Hrj.	Jmt.	Äng.	Vb.	Nb.	Ås. Lpm.	Ly. Lpm.	P. Lpm.	Lu. Lpm.	T. Lpm.
37	●	●	●	●	●	●		●	●																			
38															●		●	●	●	●	●	●	●	●	●	●	●	●
39	●	●	●	●	●	●	●	●	●	●	●	●	●	●	●		●	●	●	●	●	●	●	●	●	●	●	●
40	●	●	●	●	●	●	●	●	●	●	●	●	●	●	●	●	●	●	●	●	●	●	●	●	●	●	●	●
41	●	●	●	●		●	●	●	●	●	●	●	●	●	●	●	●	●	●	●	●	●	●	●		●	●	●
42	●	●	●	●	●	●	●	●	●	●	●	●	●	●	●	●	●	●	●	●	●	●	●	●	●	●	●	●
43																												
44													●	●			●	●		●				●	●	●	●	●
45	●	●	●	●	●	●	●	●	●	●	●	●	●	●	●	●	●	●		●		●		●	●	●	●	●
46	●	●	●	●	●	●	●	●	●	●	●	●	●	●	●	●	●	●		●		●		●	●	●	●	●
47																										●	●	●
48																												
49															●		●	●		●	●			●	●	●	●	●
50	●	●	●	●	●	●	●	●		●		●	●	●	●	●	●	●	●	●	●	●	●	●	●	●	●	●
51	●	●	●	●			●		●	●	●	●	●	●	●	●	●	●	●	●	●	●	●	●	●	●	●	
52	●	●	●	●			●		●	●	●	●	●	●	●	●	●	●	●	●	●	●	●	●	●	●	●	
53																												
54	●	●	●	●	●	●	●	●	●	●	●	●	●	●	●	●	●	●	●	●	●	●	●	●	●	●	●	●
55	●	●	●	●	●	●	●	●	●	●	●	●	●	●	●	●	●	●	●	●	●	●	●	●	●	●	●	●
56					●						●	●																
57	●	●	●	●			●	●	●	●		●	●	●	●	●					●	●	●		●			
58	●		●	●			●	●													●	●	●					
59		●												●							●					●		
60																												
61											●																	
62	●	●	●	●			●	●	●		●	●	●	●		●	●		●	●	●	●	●			●		
63	●		●	●		●	●			●																		
64	●		●	●																								
65	●	●	●				●	●																				
66		●	●	●	●	●	●	●			●	●	●	●	●	●												
67	●	●	●	●	●	●		●	●	●											●		●					
68												●	●				●	●		●	●	●	●	●	●	●	●	●
69																				●				●	●	●		
70	●			●			●																					
71																												
72	●	●	●	●		●	●	●	●	●	●	●	●	●	●	●		●	●	●	●	●	●	●	●	●	●	●
73	●	●	●	●	●	●	●	●	●	●	●	●	●	●	●													
74	●	●	●	●		●	●	●	●	●	●	●	●	●			●		●	●	●	●	●	●	●	●	●	●

211

		Ø + AK	HE (s+n)	O (s+n)	B (ø+v)	VE	TE (y+i)	AA (y+i)	VA (y+i)	R (y+i)	HO (y+i)	SF (y+i)	MR (y+i)	ST (y+i)	NT (y+i)	Ns (y+i)
Nebria salina (Fairm. & Lab.)	37							●	●	●	●	●	●	●		
Pelophila borealis (Payk.)	38	●	●	●	●		●				●	●	●	●	●	●
Notiophilus aestuans Mtsch.	39	●			●		●	●								
N. aquaticus (L.)	40	●	●	●	●	●	●	●	●	●	●	●	●	●	●	●
N. palustris (Dft.)	41	●	●		●	●	●	●	●	●	●	●	●			
N. germinyi Fauv.	42	●	●	●	●	●	●	●	●	●	●	●	●	●	●	●
N. rufipes Curt.	43															
N. reitteri Spaeth	44		●	●										●	●	●
N. biguttatus (F.)	45	●	●	●	●	●	●	●	●	●	●	●	●	●	●	●
Blethisa multipunctata (L.)	46	●	●	●	●	●	●	●	●	●				●	●	●
Diacheila arctica (Gyll.)	47															
D. polita (Fldm.)	48															
Elaphrus lapponicus Gyll.	49		●	●	●		●				●	●				●
E. uliginosus F.	50	●	●	●	●	●	●	●	●	●				●	●	
E. cupreus Dft.	51	●	●	●	●	●	●	●		●	●	●	●	●	●	
E. riparius (L.)	52	●	●	●	●	●	●	●	●		●	●	●	●	●	
E. angusticollis F. Sahlbg.	53															
Loricera pilicornis (F.)	54	●	●	●	●	●	●	●	●	●	●	●	●	●	●	●
Clivina fossor (L.)	55	●	●	●	●	●	●	●	●	●	●	●	●	●	●	●
C. collaris (Hbst.)	56															
Dyschirius thoracicus (Rossi)	57	●				●	●	●	●		●	●				●
D. obscurus (Gyll.)	58										●					
D. angustatus (Ahr.)	59			●										●	●	●
D. nitidus (Dej.)	60															
D. neresheimeri Wagn.	61															
D. politus (Dej.)	62	●	●	●	●	●	●	●			●					
D. impunctipennis Daws.	63										●					
D. chalceus Er.	64															
D. salinus Schaum	65	●					●	●								
D. aeneus (Dej.)	66	●			●		●									
D. luedersi Wagn.	67	●				●	●		●							
D. septentrionum Munst.	68		●	●										●	●	●
D. nigricornis Mtsch.	69		●													
D. intermedius Putz.	70															
D. laeviusculus Putz.	71															
D. globosus (Hbst.)	72	●	●	●	●	●	●	●	●	●	●	●	●	●	●	●
Broscus cephalotes (L.)	73	●	●		●	●		●	●					●		
Miscodera arctica (Payk.)	74	●	●	●	●			●	●	●	●	●		●		●

212

	Nn (ø+v)	TR (y+i)	F (v+i)	F (n+ø)	Al	Ab	N	Ka	St	Ta	Sa	Öa	Tb	Sb	Kb	Om	Ok	Ob S	Ob N	Ks	LkW	LkE	Le	Li	Vib	Kr	Lr
37																											
38	●	●	●	●						●	●			●	●	●	●	●	●	●	●	●	●	●	●	●	●
39							●	●	●	●	●	●	●												●	●	
40	●	●		●	●	●	●	●	●	●	●	●	●	●	●	●	●	●	●	●	●	●			●	●	
41					●	●	●	●	●	●	●	●	●	●	●	●	●	●	●	●	●	●	●		●	●	
42	●	●	●	●	●	●	●	●	●	●	●	●	●	●	●	●	●	●	●	●	●	●	●		●	●	
43																											
44			●	●	●	●							●		●												
45	●	●		●	●	●	●	●	●	●	●	●	●	●	●	●	●	●	●	●	●	●	●	●	●	●	
46	●	●		●	●	●	●	●	●	●	●	●	●	●	●	●	●	●	●	●	●	●	●	●	●	●	
47			●	●																	●	●	●				●
48																											●
49	●	●	●	●							●					●					●	●	●				
50					●	●	●	●	●	●	●	●	●	●	●	●	●	●	●	●	●	●	●				
51	●	●	●	●	●	●	●	●	●	●	●	●	●	●	●	●	●	●	●	●	●	●	●				
52	●	●	●	●	●	●	●	●	●	●	●	●	●	●	●	●	●	●	●	●	●	●	●				
53																									●	●	
54	●	●	●	●	●	●	●	●	●	●	●	●	●	●	●	●	●	●	●	●	●	●	●		●	●	
55	●	●	●	●	●	●	●	●	●	●	●	●	●	●	●	●	●	●	●	●	●	●	●		●	●	
56																											
57					●	●	●	●	●	●	●	●	●	●	●	●	●	●	●	●					●	●	●
58						●	●	●			●				●			●		●					●	●	●
59		●	●																						●		●
60						●				●																	
61																											
62					●	●	●			●	●	●	●	●	●			●				●			●	●	
63						●			●						●		●								●	●	
64																											
65					●	●	●																				
66					●	●	●									●										●	
67					●	●	●	●	●	●	●	●	●				●		●						●	●	●
68	●	●	●	●				●			●	●	●		●	●	●		●	●	●	●	●	●	●	●	●
69		●	●																●	●		●	●				●
70																										●	
71																											
72	●	●	●	●	●	●	●	●		●	●	●	●	●	●	●	●	●		●	●	●	●	●			
73					●	●	●	●	●	●	●	●	●	●	●	●	●		●	●	●	●			●	●	●
74	●	●	●	●	●	●	●	●	●	●	●	●													●	●	●

		Germany	G. Britain	SJ	EJ	WJ	NWJ	NEJ	F	LFM	SZ	NWZ	NEZ	B	Sk.	Bl.
Patrobus septentrionis Dej.	75	●	●													
P. australis J. Sahlbg.	76	●		●					●	●	●			●	●	
P. assimilis Chaud.	77	●	●		●	●							●		●	●
P. atrorufus (Ström)	78	●	●	●	●	●	●	●	●	●	●	●	●	●	●	●
Perileptus areolatus (Creutz.)	79	●														
Aepus marinus (Ström)	80	●														
Trechus secalis (Payk.)	81	●	●	●	●	●	●	●	●	●	●	●	●	●	●	●
T. rivularis (Gyll.)	82	●	●								●					●
T. rubens (F.)	83	●	●	●	●		●				●			●	●	●
T. fulvus Dej.	84	●	●													
T. quadristriatus (Schrk.)	85	●	●	●	●	●	●	●	●	●	●	●	●	●	●	●
T. obtusus Er.	86	●	●	●	●	●	●		●	●	●	●	●	●	●	●
T. micros (Hbst.)	87	●	●	●	●	●		●	●	●	●	●	●	●	●	●
T. discus (F.)	88	●	●	●	●	●		●	●	●	●	●	●	●	●	●
Asaphidion pallipes (Dft.)	89	●	●	●	●	●	●	●	●		●	●	●	●	●	●
A. flavipes (L.)	90	●	●	●	●	●		●	●	●	●	●	●	●	●	●
A. curtum Heyd.	91	●				●			●	●	●	●	●			
Bembidion striatum (F.)	92	●									●					
B. velox (L.)	93	●													●	
B. lapponicum Zett.	94															
B. litorale (Oliv.)	95	●	●	●	●	●					●					
B. argenteolum Ahr.	96	●														
B. nigricorne Gyll.	97	●	●			●	●		●							
B. pygmaeum (F.)	98	●														
B. lampros (Hbst.)	99	●	●	●	●	●	●	●	●	●	●	●	●	●	●	●
B. properans (Stph.)	100	●	●	●	●	●	●	●	●	●	●	●	●	●	●	●
B. obtusum Aud.-Serv.	101	●	●	●	●	●	●	●	●	●	●	●	●	●	●	●
B. harpaloides Aud.-Serv.	102	●	●													
B. quinquestriatum Gyll.	103	●	●	●	●	●			●		●	●		●	●	●
B. punctulatum Drap.	104	●	●													
B. bipunctatum (L.)	105	●	●	●	●	●	●	●			●		●	●	●	●
B. ruficolle (Pz.)	106	●									●			●	●	●
B. pallidipenne (Ill.)	107	●	●	●	●	●	●	●			●				●	●
B. dentellum (Thbg.)	108	●	●								●				●	●
B. tinctum Zett.	109															
B. varium (Oliv.)	110	●	●	●	●	●	●	●	●	●	●	●	●	●	●	●
B. semipunctatum (Don.)	111	●	●												●	●
B. obliquum Sturm	112	●	●	●	●	●	●	●	●	●	●	●	●	●	●	●

214

	Hall.	Sm.	Öl.	Gtl.	G. Sand.	Ög.	Vg.	Boh.	Dlsl.	Nrk.	Sdm.	Upl.	Vstm.	Vrm.	Dlr.	Gstr.	Hls.	Med.	Hrj.	Jmt.	Ång.	Vb.	Nb.	Ås. Lpm.	Ly. Lpm.	P. Lpm.	Lu. Lpm.	T. Lpm.
75															●		●	●	●	●	●	●	●	●	●	●	●	●
76																												
77	●	●	●	●		●	●	●	●	●	●	●	●	●	●	●	●	●	●	●	●	●	●	●	●	●	●	●
78	●	●	●	●		●	●	●	●	●	●	●	●	●	●	●	●			●	●	●	●	●			●	
79	●												●	●		●												
80								●																				
81	●	●	●	●		●	●	●	●	●	●	●	●	●	●	●	●			●	●	●	●	●				
82		●				●	●	●	●	●	●	●	●	●		●					●						●	
83	●	●	●	●		●	●	●	●	●	●	●	●	●	●	●	●											
84																												
85	●	●	●	●	●	●	●	●	●	●	●	●	●	●	●	●	●	●		●	●		●					
86		●	●	●															●	●				●	●			
87	●						●	●	●							●				●	●							
88	●			●			●	●	●							●												
89		●	●	●			●	●	●			●		●						●	●	●	●	●			●	
90	●	●	●	●	●	●	●	●	●	●	●	●	●	●	●	●	●	●										
91																												
92																												
93	●	●		●	●		●	●	●	●			●	●	●	●	●	●	●	●	●	●	●		●	●		●
94																			●	●							●	●
95		●					●							●				●	●		●	●	●					
96													●	●			●											
97	●	●				●	●			●								●				●						
98																												
99	●	●	●	●		●	●	●	●	●	●	●	●	●	●	●	●	●	●	●	●	●	●	●	●	●	●	
100	●	●	●	●		●	●	●	●	●	●	●		●	●													
101	●	●	●	●			●	●	●		●																	
102																												
103	●	●	●	●			●	●						●														
104																												
105	●	●	●	●		●	●	●	●			●	●	●	●	●	●	●	●	●	●	●	●	●	●	●	●	●
106		●		●	●									●	●			●										
107	●	●	●	●			●																					
108	●	●	●	●			●	●	●	●	●	●	●	●	●					●	●	●						
109																●				●	●	●	●	●			●	●
110	●	●	●	●	●	●	●			●	●	●								●								
111			●											●	●													
112	●	●	●	●	●	●	●	●	●	●	●	●	●	●	●	●	●	●	●	●	●	●	●	●			●	●

NORWAY

		Ø+AK	HE (s+n)	O (s+n)	B (ø+v)	VE	TE (y+i)	AA (y+i)	VA (y+i)	R (y+i)	HO (y+i)	SF (y+i)	MR (y+i)	ST (y+i)	NT (y+i)	Ns (y+i)
Patrobus septentrionis Dej.	75	●	●	●	●		●	●		●	●	●	●	●	●	●
P. australis J. Sahlbg.	76															
P. assimilis Chaud.	77	●	●	●	●		●	●	●	●	●	●	●	●	●	●
P. atrorufus (Ström)	78	●	●	●	●		●	●	●	●	●	●	●	●	●	●
Perileptus areolatus (Creutz.)	79	●	●		●	●	●		●					●		
Aepus marinus (Ström)	80	●									●	●	●	●		
Trechus secalis (Payk.)	81	●	●	●	●	●	●	●	●	●	●	●	●	●	●	●
T. rivularis (Gyll.)	82	●												●		
T. rubens (F.)	83	●	●	●	●	●	●	●	●	●	●	●	●	●	●	●
T. fulvus Dej.	84								●				●			
T. quadristriatus (Schrk.)	85	●	●	●	●	●	●	●	●	●						
T. obtusus Er.	86										●	●	●	●	●	●
T. micros (Hbst.)	87	●	●		●	●								●		
T. discus (F.)	88	●			●									●		
Asaphidion pallipes (Dft.)	89	●	●	●	●		●						●	●		●
A. flavipes (L.)	90	●	●		●	●										
A. curtum Heyd.	91															
Bembidion striatum (F.)	92															
B. velox (L.)	93	●	●	●	●		●	●	●	●				●	●	
B. lapponicum Zett.	94													●	●	●
B. litorale (Oliv.)	95	●	●											●	●	
B. argenteolum Ahr.	96			●	●									●	●	
B. nigricorne Gyll.	97	●														
B. pygmaeum (F.)	98															
B. lampros (Hbst.)	99	●	●	●	●	●	●	●	●	●	●	●	●	●	●	●
B. properans (Stph.)	100	●	●	●	●	●	●	●								
B. obtusum Aud.-Serv.	101															
B. harpaloides Aud.-Serv.	102								●							
B. quinquestriatum Gyll.	103	●			●											
B. punctulatum Drap.	104															
B. bipunctatum (L.)	105	●	●	●	●	●	●	●	●	●	●	●	●	●	●	●
B. ruficolle (Pz.)	106															
B. pallidipenne (Ill.)	107	●							●	●						
B. dentellum (Thbg.)	108	●	●		●	●	●	●			●	●	●	●	●	
B. tinctum Zett.	109															
B. varium (Oliv.)	110	●														
B. semipunctatum (Don.)	111	●	●	●	●	●								●	●	
B. obliquum Sturm	112	●	●	●	●	●	●	●	●	●	●	●		●	●	

	Nn (ø+v)	TR (y+i)	F (v+i)	F (n+ø)	Al	Ab	N	Ka	St	Ta	Sa	Oa	Tb	Sb	Kb	Om	Ok	ObS	ObN	Ks	LkW	LkE	Le	Li	Vib	Kr	Lr
75	●	●	●	●			●					●		●			●	●	●				●	●		●	●
76						●			●	●	●		●	●	●											●	●
77	●	●	●	●	●	●	●	●	●	●	●	●	●	●	●	●	●	●	●	●	●	●	●	●		●	●
78	●				●	●	●	●	●	●	●	●														●	●
79																										●	
80																											
81		●			●	●	●	●	●	●	●	●	●													●	●
82					●	●	●	●	●	●	●	●	●				●	●								●	●
83	●	●	●	●	●	●	●	●	●	●	●	●	●	●	●	●	●	●	●	●	●				●	●	●
84																											
85					●	●	●	●	●	●	●	●	●	●	●											●	●
86	●	●																									
87						●	●																			●	●
88					●	●	●																			●	●
89		●	●	●		●	●	●	●	●	●	●	●	●	●	●	●	●		●					●	●	●
90					●	●	●	●	●	●	●	●	●													●	●
91																											
92						●																			●		
93		●	●	●	●	●	●	●	●	●	●	●	●	●	●	●	●	●	●	●	●	●	●	●		●	●
94		●	●	●													●	●		●	●					●	●
95								●	●	●	●	●	●	●			●	●	●			●				●	●
96																											
97						●	●		●	●	●						●	●	●							●	●
98						●																					
99					●	●	●	●	●	●	●	●	●	●	●	●	●	●	●	●	●					●	●
100					●	●	●	●	●	●	●	●	●	●	●											●	●
101																											
102																											
103																											
104							●																			●	●
105	●	●	●	●	●	●	●	●	●		●	●	●	●	●	●	●	●	●	●	●					●	●
106						●	●	●	●		●	●			●				●		●		●			●	●
107																											
108						●	●	●	●	●	●			●	●	●					●					●	●
109							●				●			●		●	●	●	●	●	●			●		●	●
110					●	●	●			●		●															
111						●																				●	●
112					●	●	●	●	●	●	●	●	●	●	●	●	●	●	●	●	●	●	●	●		●	●

			DENMARK												
		Germany	G. Britain	SJ	EJ	WJ	NWJ	NEJ	F	LFM	SZ	NWZ	NEZ	B	Sk.	Bl.
Bembidion ephippium (Marsh.)	113	●	●			●										
B. minimum (F.)	114	●	●	●	●	●	●	●	●	●	●	●	●	●	●	●
B. normannum Dej.	115	●	●	●	●	●			●	●		●				
B. azurescens (D. Torre)	116	●														
B. tenellum Er.	117	●			●				●	●	●				●	●
B. articulatum (Pz.)	118	●	●	●	●	●	●	●	●	●	●	●	●	●		●
B. octomaculatum (Gz.)	119	●	●												●	●
B. doris (Pz.)	120	●	●	●	●	●	●	●	●	●	●	●	●	●	●	●
B. schuppelii Dej.	121	●	●		●											
B. chaudoirii Chaud.	122															
B. gilvipes Sturm	123	●	●	●				●	●	●	●	●	●	●	●	●
B. fumigatum (Dft.)	124	●	●	●	●	●	●	●	●	●	●	●	●	●	●	●
B. assimile Gyll.	125	●	●	●	●	●		●	●	●	●	●	●	●	●	●
B. clarkii (Daws.)	126	●	●	●		●				●			●	●	●	●
B. transparens (Gebl.)	127	●							●							
B. quadrimaculatum (L.)	128	●	●	●	●	●		●	●	●	●		●	●	●	●
B. humerale Sturm	129	●	●	●	●	●							●	●	●	
B. quadripustulatum Aud.- Serv.	130	●	●													
B. biguttatum (F.)	131	●	●	●	●	●		●	●	●	●	●	●	●	●	●
B. iricolor Bedel	132	●	●	●	●											
B. lunulatum (Fourc.)	133	●	●	●	●			●	●	●	●	●	●	●	●	●
B. aeneum Germ.	134	●	●	●	●	●	●	●	●	●	●	●	●	●	●	●
B. guttula (F.)	135	●	●	●	●	●	●	●	●	●	●	●	●	●	●	●
B. mannerheimii Sahlbg.	136	●	●	●	●	●	●	●	●	●	●	●	●	●	●	●
B. laterale (Sam.)	137	●	●	●	●	●										
B. genei Küst.	138	●	●	●	●	●	●	●	●	●	●	●	●	●	●	●
B. fellmanni Mann.	139															
B. difficile (Mtsch.)	140															
B. crenulatum F. Sahlbg.	141															
B. prasinum (Dft.)	142	●														
B. hyperboreorum Munst.	143															
B. virens Gyll.	144	●														
B. hastii Sahlbg.	145															
B. hirmocaelum Chaud.	146															
B. tibiale (Dft.)	147	●														
B. mckinleyi Fall	148															
B. monticola Sturm	149	●			●	●										
B. nitidulum (Marsh.)	150	●	●	●	●	●	●	●	●	●	●	●	●	●	●	●

	Hall.	Sm.	Öl.	Gtl.	G. Sand.	Ög.	Vg.	Boh.	Dlsl.	Nrk.	Sdm.	Upl.	Vstm.	Vrm.	Dlr.	Gstr.	Hls.	Med.	Hrj.	Jmt.	Äng.	Vb.	Nb.	Ås. Lpm.	Ly. Lpm.	P. Lpm.	Lu. Lpm.	T. Lpm.
113																												
114	●	●	●	●	●		●	●			●																	
115																												
116				●																								
117																												
118	●	●	●	●	●	●	●	●		●	●	●	●	●	●	●												
119				●																								
120	●	●	●	●	●	●	●	●	●	●	●	●	●	●	●	●	●	●	●	●	●	●	●	●		●	●	●
121												●			●					●	●	●		●				
122																												
123	●	●	●	●	●	●	●	●	●	●	●	●	●	●	●	●		●	●		●	●						
124	●																											
125	●	●	●	●	●	●	●	●		●	●	●	●	●														
126	●	●	●																									
127	●	●				●	●																					
128	●	●	●	●	●	●	●	●	●	●	●	●	●	●	●	●	●	●	●	●	●	●	●		●	●		
129	●	●		●			●					●																
130																												
131					●																							
132																												
133		●																										
134	●	●	●	●	●	●	●	●	●	●		●	●	●	●		●						●					
135	●	●	●	●	●	●	●	●	●	●	●	●	●	●	●	●		●	●	●	●	●	●					
136	●	●	●												●				●	●								
137																												
138	●	●	●	●		●	●	●	●	●	●	●	●															
139																●			●	●						●	●	●
140											●	●				●	●	●	●	●	●	●	●	●		●	●	●
141																												
142											●			●	●	●	●	●	●	●	●	●	●	●			●	●
143																	●					●	●	●			●	●
144					●						●			●	●	●	●	●	●	●	●	●	●	●			●	●
145																		●	●		●	●	●	●	●	●	●	●
146																												
147																												
148																												●
149																												
150	●	●				●	●	●	●			●								●				●	●			

219

NORWAY

		Ø+AK	HE (s+n)	O (s+n)	B (ø+v)	VE	TE (y+i)	AA (y+i)	VA (y+i)	R (y+i)	HO (y+i)	SF (y+i)	MR (y+i)	ST (y+i)	NT (y+i)	Ns (y+i)
Bembidion ephippium (Marsh.)	113															
B. minimum (F.)	114	●			●		●	●						●	●	
B. normannum Dej.	115															
B. azurescens (D. Torre)	116															
B. tenellum Er.	117															
B. articulatum (Pz.)	118	●	●		●	●	●	●								
B. octomaculatum (Gz.)	119															
B. doris (Pz.)	120	●	●	●	●	●	●	●	●	●	●			●	●	●
B. schuppelii Dej.	121		●	●										●	●	●
B. chaudoirii Chaud.	122															
B. gilvipes Sturm	123	●	●	●	●	●										
B. fumigatum (Dft.)	124															
B. assimile Gyll.	125	●				●	●		●	●						
B. clarkii (Daws.)	126															
B. transparens (Gebl.)	127															
B. quadrimaculatum (L.)	128	●	●	●	●	●	●	●	●					●	●	●
B. humerale Sturm	129															
B. quadripustulatum Aud.- Serv.	130															
B. biguttatum (F.)	131															
B. iricolor Bedel	132															
B. lunulatum (Fourc.)	133															
B. aeneum Germ.	134	●			●									●	●	●
B. guttula (F.)	135	●	●	●	●	●	●	●	●							
B. mannerheimii Sahlbg.	136	●	●	○	●	●	●	●	●	●						
B. laterale (Sam.)	137															
B. genei Küst.	138															
B. fellmanni Mann.	139		●	●	●				●	●	●			●	●	●
B. difficile (Mtsch.)	140		●	●	●							●	●			●
B. crenulatum F. Sahlbg.	141															
B. prasinum (Dft.)	142	●	●	●	●		●	●	●	●	●			●	●	●
B. hyperboraeorum Munst.	143															●
B. virens Gyll.	144	●	●	●	●	●	●	●	●	●	●			●	●	●
B. hastii Sahlbg.	145		●	●					●	●	●			●	●	●
B. hirmocaelum Chaud.	146															
B. tibiale (Dft.)	147								●							
B. mckinleyi Fall	148															
B. monticola Sturm	149															
B. nitidulum (Marsh.)	150	●	●	●	●	●	●	●	●	●	●	●	●	●	●	●

	Nn (ø+v)	TR (y+i)	F (v+i)	F (n+ø)	Al	Ab	N	Ka	St	Ta	Sa	Öa	Tb	Sb	Kb	Öm	Ok	Ob S	Ob N	Ks	LkW	LkE	Lc	Li	Vib	Kr	Lr
113																											
114					●	●	●					●													●		
115																											
116																									●	●	
117																									●		
118					●	●	●	●	●	●	●				●										●	●	
119							●																				
120		●			●	●	●	●	●	●	●	●	●	●	●	●	●	●	●	●	●	●	●	●	●	●	●
121	●	●	●			●	●							●	●		●	●	●	●			●		●	●	●
122																									●		
123					●	●	●	●	●	●	●	●	●	●	●	●									●	●	
124																											
125					●		●																				
126																											
127	●	●			●	●	●	●	●	●	●		●		●		●	●	●		●		●		●	●	●
128		●	●		●	●	●	●	●	●	●	●	●	●	●	●	●	●	●	●	●	●	●	●	●	●	●
129					●	●	●		●	●					●										●	●	
130																											
131						●	●		●	●															●		
132																											
133																											
134	●																									●	
135					●	●	●	●	●	●	●	●	●	●	●	●	●			●					●	●	●
136					●	●	●	●	●	●	●	●	●	●	●	●									●	●	
137																											
138					●		●																				
139		●	●	●																●	●		●	●	●	●	
140	●	●	●	●													●		●	●	●	●	●		●	●	
141																											●
142	●	●	●	●																●	●	●	●	●		●	●
143	●	●	●													●					●	●		●		●	●
144	●	●	●														●		●		●	●		●		●	●
145	●	●	●														●	●	●	●	●	●				●	●
146																										●	
147																											
148	●	●	●																				●				
149						●									●												
150								●			●				●										●	●	

221

				Germany	G. Britain	SJ	EJ	WJ	NWJ	NEJ	F	LFM	SZ	NWZ	NEZ	B	Sk.	Bl.
								DENMARK										
Bembidion grapii Gyll.	151																	
B. yukonum Fall	152																	
B. dauricum (Mtsch.)	153																	
B. stephensi Crotch	154				●	●	●	●		●		●	●	●		●	●	●	
B. lunatum (Dft.)	155				●	●	●	●	●	●		●						●	●
B. tetracolum Say	156				●	●	●	●	●	●	●	●	●	●	●	●	●	●	●
B. bruxellense Wesm.	157				●	●	●	●	●	●	●	●	●	●	●	●	●	●	●
B. petrosum Gebl.	158																		
B. andreae (F.)	159				●	●	●	●				●	●	●		●	●	●	●
B. femoratum Sturm	160				●	●	●	●	●	●	●	●	●	●	●	●	●	●	●
B. maritimum (Stph.)	161				●	●	●			●									
B. obscurellum (Mtsch.)	162									●		●							
B. saxatile Gyll.	163				●	●	●	●		●		●	●	●	●	●	●	●	●
B. decorum (Zenk.)	164				●	●													
Tachys bistriatus (Dft.)	165				●	●			●			●					●		
T. bisulcatus (Nic.)	166				●	●	●	●				●	●			●		●	●
T. parvulus (Dej.)	167				●														
Tachyta nana (Gyll.)	168				●													●	
Pogonus luridipennis (Germ.)	169				●	●	●			●									
P. chalceus (Marsh.)	170				●	●	●	●		●		●							

	Hall.	Sm.	Öl.	Gtl.	G. Sand.	Ög.	Vg.	Boh.	Dlsl.	Nrk.	Sdm.	Upl.	Vstm.	Vrm.	Dlr.	Gstr.	Hls.	Med.	Hrj.	Jmt.	Ång.	Vb.	Nb.	Ås. Lpm.	Ly. Lpm.	P. Lpm.	Lu. Lpm.	T. Lpm.
151														●	●		●	●	●	●	●	●	●	●	●	●	●	●
152																												●
153																											●	●
154							●																					
155			●				●							●				●		●								
156	●	●	●	●		●	●	●	●	●	●	●	●	●	●		●	●										
157	●	●	●	●		●	●	●	●	●	●	●	●	●	●	●	●	●	●	●	●	●	●	●	●	●	●	●
158														●														●
159	●			●																								
160	●	●	●	●		●	●	●		●	●	●	●	●	●	●	●	●	●	●	●	●			●	●		
161																												
162			●	●																								
163	●	●	●	●		●	●	●	●	●	●	●	●	●	●	●	●	●	●	●	●	●	●				●	
164																												
165																												
166	●	●	●	●	●	●	●			●		●		●	●	●	●			●	●	●		●	●	●	●	
167	●																											
168		●				●				●	●	●	●	●	●	●	●			●	●	●	●	●	●	●		
169							●																					
170																												

223

		Ø+AK	HE (s+n)	O (s+n)	B (ø+v)	VE	TE (y+i)	AA (y+i)	VA (y+i)	R (y+i)	HO (y+i)	SF (y+i)	MR (y+i)	ST (y+i)	NT (y+i)	Ns (y+i)
Bembidion grapii Gyll.	151	●	●						●	●				●		●
B. yukonum Fall	152															
B. dauricum (Mtsch.)	153										●					
B. stephensi Crotch	154	●														
B. lunatum (Dft.)	155	●	●	●	●	●							●	●	●	●
B. tetracolum Say	156	●	●		●	●	●	●	●	●	●	●	●	●	●	
B. bruxellense Wesm.	157	●	●		●	●	●	●	●	●	●	●	●	●	●	●
B. petrosum Gebl.	158	●	●										●	●	●	●
B. andreae (F.)	159															
B. femoratum Sturm	160	●	●	●	●	●	●	●	●	●	●	●	●	●	●	●
B. maritimum (Stph.)	161															
B. obscurellum (Mtsch.)	162															
B. saxatile Gyll.	163	●	●	●	●	●	●	●	●	●	●	●	●	●	●	●
B. decorum (Zenk.)	164															
Tachys bistriatus (Dft.)	165															
T. bisulcatus (Nic.)	166	●	●				●								●	
T. parvulus (Dej.)	167															
Tachyta nana (Gyll.)	168	●	●		●		●	●								
Pogonus luridipennis (Germ.)	169	●														
P. chalceus (Marsh.)	170															

	Nn (ø+v)	TR (y+i)	F (v+i)	F (n+ø)	Al	Ab	N	Ka	St	Ta	Sa	Öa	Tb	Sb	Kb	Om	Ok	ObS	ObN	Ks	LkW	LkE	Lc	Li	Vib	Kr	Lr
151	●	●	●	●	●	●	●		●				●	●	●	●	●		●	●	●	●		●	●	●	●
152		●	●																					●			●
153	●																				●						
154								●	●		●														●		
155																											
156					●	●	●	●	●	●	●	●	●		●	●								●	●		
157	●	●	●	●	●	●	●	●	●	●	●	●	●	●	●	●	●	●	●	●	●	●	●	●	●	●	●
158		●	●	●																●				●			
159					●	●	●	●																●	●		
160	●	●	●	●		●	●	●	●	●	●		●	●	●		●					●		●	●		
161																											
162																											●
163	●	●	●		●	●	●	●	●	●	●	●		●	●	●			●	●		●		●	●	●	●
164																											
165											●														●		
166							●	●	●	●	●		●		●		●							●	●	●	
167																											
168		●		●	●	●	●	●	●	●	●	●	●	●	●	●	●	●	●	●	●	●		●	●	●	●
169																											
170																											

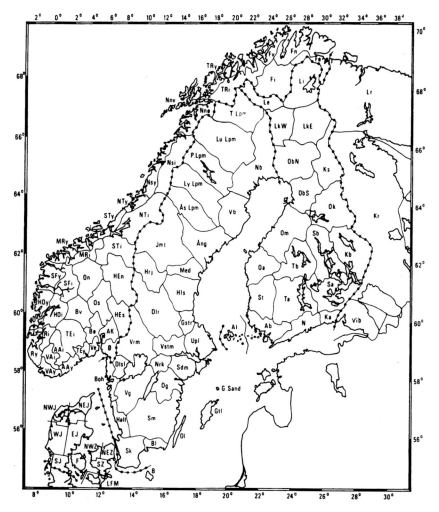

List of abbreviations for the provinces used throughout the text, on the map and in the following tables.

DENMARK

SJ	South Jutland	LFM	Lolland, Falster, Møn
EJ	East Jutland	SZ	South Zealand
WJ	West Jutland	NWZ	North West Zealand
NWJ	North West Jutland	NEZ	North East Zealand
NEJ	North East Jutland	B	Bornholm
F	Funen		

SWEDEN

Sk.	Skåne	Vrm.	Värmland
Bl.	Blekinge	Dlr.	Dalarna
Hall.	Halland	Gstr.	Gästrikland
Sm.	Småland	Hls.	Hälsingland
Öl.	Öland	Med.	Medelpad
Gtl.	Gotland	Hrj.	Härjedalen
G. Sand.	Gotska Sandön	Jmt.	Jämtland
Ög.	Östergötland	Äng.	Ångermanland
Vg.	Västergötland	Vb.	Västerbotten
Boh.	Bohuslän	Nb.	Norrbotten
Dlsl.	Dalsland	Äs. Lpm.	Äsele Lappmark
Nrk.	Närke	Ly. Lpm.	Lycksele Lappmark
Sdm.	Södermanland	P. Lpm.	Pite Lappmark
Upl.	Uppland	Lu. Lpm.	Lule Lappmark
Vstm.	Västmanland	T. Lpm.	Torne Lappmark

NORWAY

Ø	Østfold	HO	Hordaland
AK	Akershus	SF	Sogn og Fjordane
HE	Hedmark	MR	Møre og Romsdal
O	Opland	ST	Sør-Trøndelag
B	Buskerud	NT	Nord-Trøndelag
VE	Vestfold	Ns	southern Nordland
TE	Telemark	Nn	northern Nordland
AA	Aust-Agder	TR	Troms
VA	Vest-Agder	F	Finnmark
R	Rogaland		

n northern   s southern   ø eastern   v western   y outer   i inner

FINLAND

Al	Alandia	Kb	Karelia borealis
Ab	Regio aboensis	Om	Ostrobottnia media
N	Nylandia	Ok	Ostrobottnia kajanensis
Ka	Karelia australis	ObS	Ostrobottnia borealis, S part
St	Satakunta	ObN	Ostrobottnia borealis, N part
Ta	Tavastia australis	Ks	Kuusamo
Sa	Savonia australis	LkW	Lapponia kemensis, W part
Oa	Ostrobottnia australis	LkE	Lapponia kemensis, E part
Tb	Tavastia borealis	Li	Lapponia inarensis
Sb	Savonia borealis	Le	Lapponia enontekiensis

USSR

Vib   Regio Viburgensis   Kr   Karelia rossica   Lr   Lapponia rossica

Printed in the United States
By Bookmasters